The Patrick Moore Practical Astronomy Series

Series Editor

Gerald R. Hubbell
Mark Slade Remote Observatory, Locust Grove, VA, USA

W0235090

More information about this series at http://www.springer.com/series/3192

The stars are friends of mine; to lofty height
When falls the sombre canopy of night
Upon a slumbering world, my spirit flies
And treads with them the highways of the skies
I leap from world to world while they rehearse
The mighty chorus of the Universe
I explore skies of azure blue,
Sprinkled with diamonds of varied hue
Seek the lost Pleiad through skies aflame
And learn from her the secret of her shame.
Stars mark the ways of men and shake with mirth
At all the customs of this lowly Earth
Great wisdom and great mysteries they know
They tell the tales of long ago,
Ere time was born and chaos had its way
And darkness held its mantle over day
Why should I prize the boasted things of Earth
when I can walk with stars and share their mirth?
To know their wonders is to be divine
I'd rather walk with stars; they're friends of mine.

- Richard Mann

Observer's Guide to Variable Stars

Martin Griffiths

 Springer

Martin Griffiths
Dark Sky Wales, Blackmill, Bridgend
Wales, UK

ISSN 1431-9756   ISSN 2197-6562   (electronic)
The Patrick Moore Practical Astronomy Series
ISBN 978-3-030-00903-8   ISBN 978-3-030-00904-5   (eBook)
https://doi.org/10.1007/978-3-030-00904-5

Library of Congress Control Number: 2018957655

Image Credit: NASA

This Springer imprint is published by the registered company Springer Nature Switzerland AG
The registered company address is: Gewerbestrasse 11, 6330 Cham, Switzerland

This book is dedicated to my wonderful wife Dena with thanks for all your support and love and to our mad dogs Gloria, Ianto and Jango that drive us crazy.

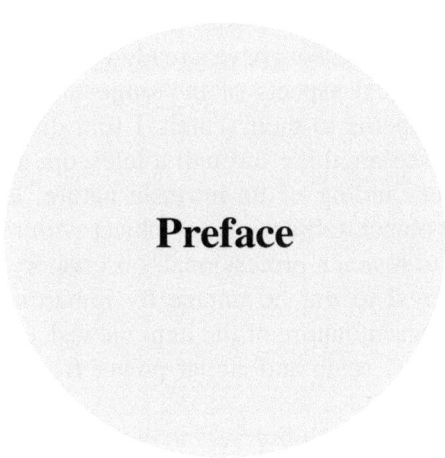

Preface

Variable stars are one of the most diverse and rewarding fields of study in astronomy. Such stars not only vary in brightness, but understanding the nature of this variability leads us to know the most amazing and wonderful sets of objects and a deeper knowledge of stellar systems and evolution. In short, variable stars are fascinating, beautiful and enchanting.

I have been an astronomer from a young age and have been captivated by many objects that have become personal favourites; as with most other astronomers, one returns time and again to old familiar ones no matter how often they have been examined in the past. Variable stars provide an observer with such a range of objects and a diversity of forms that the observer will always find something new to occupy them.

I have set this book out in a format that will hopefully enable the novice to pick up the information and go with it into the observing field. I do not expect everyone to engage completely with the astrophysical concepts, although learning about such adds stimulus to variable star research, as I believe that a good understanding of the processes involved can add to the observing experience. This field of research is also one in which individual discoveries can still be made, and the contribution of amateur astronomers is often all that we have as a scientific community on the nature and processes behind such stars. It is a field where the mundane is anything but that.

I hope that this volume will provide the tools necessary to start searching for these wonderful entities. It does not matter what the aperture of your telescope is or how frequently you observe. The stars included in this book I hope will please and delight most observers. I have attempted to strike a balance between easily visible objects that can be seen in any telescope or binoculars and variable stars that are a direct challenge to those with large aperture equipment or access to photometric tools and methods.

I have also attempted to cover a brief historical and physical analysis of variable stars in order that readers have a ready volume covering both observational and astrophysical aspects of the subject, which will give added understanding and impetus to their search. I find that when teaching students, the ability to see anything through a telescope is augmented by the fuller physical understanding of its intrinsic nature, leading to a greater appreciation for the object. Observing any object with relatively small telescopes is not going to reveal a professional, observatory quality image. But this lack can be turned to our advantage by imparting some foreground knowledge on the inherent nature of the item viewed, enabling an appreciation of cosmic distance, scale and stellar power from our fleeting earthly platform.

In the final analysis, I want observers to enjoy their experiences in hunting down these wonderful stars and discovering them for themselves. I hope this small book will help one to grow in knowledge and appreciation of one striking facet of the universe around us.

Dark Sky Wales, Blackmill, Bridgend, Wales, UK   Martin Griffiths
2018

Acknowledgements

I have been looking at the stars since I was a young boy and have spent many nights marvelling at the wonders of the night sky. As I moved from amateur to professional astronomy, my love of the sky and my amazement at all it contains have never dimmed. Along the way, I have encountered many people who have encouraged me and have helped make my dreams come true, not least of which are those who encouraged me to write. This is the sixth book that I have written and one which I have enjoyed researching and writing. As an astronomer, I am keenly aware of the importance of variable stars and their application to so many fields of astronomical science.

I would like to thank the staff at Springer for their help and encouragement and also the helpful staff at the BAA Variable Star Section, especially Roger Pickard, to whom I am indebted for the images of star fields and comparisons for the final chapter of the book. In addition, the AAVSO staff and observers have proved to be very helpful in offering advice and allowing me to draw upon their expertise and online materials to provide examples and to illustrate how much help is available to variable star observers should they take up this field of research.

Unless otherwise indicated in the text, all photographs are copyrighted.

Contents

Chapter 1

An Introduction to Variable Stars

Monitoring and recording variable stars is one of the oldest and noblest activities in amateur astronomy. In very few other fields is it possible for the modestly equipped observer to make discoveries of extreme significance and enable professional astronomers to follow up on the astrophysical aspects of such phenomena.

We have learned an enormous amount from watching the differences in light output from such stars. Although the heavens were seen to be immutable and unchanging for millennia, the discovery of variable stars showed that the ancient ideas were incorrect. Variable stars ushered in a new era in science as their true nature demanded the application of various disciplines from physics and mathematics, from chemistry to spectroscopy and to photography, photometry and cartography.

With this in mind, it would seem that modern variable star observation by amateurs is redundant. However, nothing could be further from the truth. There are a huge range of large instrumental surveys of variable stars, but it must be remembered that they do not provide the same coverage that visual observers historically have. In addition, very few surveys fully cover the same brightness range available to visual observers, and many surveys are from a single location and are dependent on weather conditions and other factors. Having a host of observers worldwide covering many objects and overlapping some provides adequate coverage, a sense of purpose and the bonds of sharing something special together.

© Springer Nature Switzerland AG 2018   1
M. Griffiths, *Observer's Guide to Variable Stars*, The Patrick Moore
Practical Astronomy Series, https://doi.org/10.1007/978-3-030-00904-5_1

What are variable stars? How were they discovered and what distinct types of variable stars are there? As one reads through this book it may seem to be a mindstorm of letters, abbreviations and catalogues, but remember that we are standing at the end of over 400 years of astronomical discovery. Let us examine the history of discovery and then turn to a brief overview of their types.

History of Variable Stars

According to S. Porceddu et al writing in the *Cambridge Journal of Archaeology* in 2008 andagain in 2013, there exists an ancient Egyptian calendar comprised of lucky and unlucky days. Exactly what this means is not relevant here, but this calendar, composed some 3,200 years ago, contains some interesting astronomical information, and these scholars suggest that it may be the oldest preserved historical document of the discovery of a variable star. The star is the eclipsing binary Algol in the constellation of Perseus. We know that the Egyptians, among many other ancient civilizations, were avid watchers of the sky, so such an observation may well be possible. Indeed, the Persian astronomer Al Sufi in his *Book of the Fixed Stars* mentions the possibility of Algol's variability in the year A. D. 964.

The first definitive variable star was recorded in 1638 when Johan Holwarda noticed that o Ceti (later named Mira) pulsated in a cycle taking about 11 months. However, the star had previously been discovered by David Fabricius in 1596, but he thought that it was a nova. Holwarda's discovery, combined with the earlier supernova of Tycho Brahe in 1572 and that of Johannes Kepler in 1604, were ground breaking in that they proved that the stars were not invariable as Aristotle and other ancient philosophers had taught. In this way, the discovery of variable stars contributed to the astronomical revolution of the sixteenth and early seventeenth centuries.

The second variable star to be described was the eclipsing variable Algol, by Geminiaro Montanari in 1669, but it was left to John Goodricke to present the correct explanation of its variability in 1784. The long period variable χ Cygni was identified in 1686 by the astronomer Gottfried Kirch, then the star R Hydrae in 1704 by Dominico Maraldi.

To understand the variability of such stars, astronomers had to observe their variability over specified time periods and draw a graph showing the differences in brightness over time. Such graphs are known as light curves. In the study of objects that change their brightness over time, such as novae, supernovae, and variable stars, the light curve is a simple but valuable tool

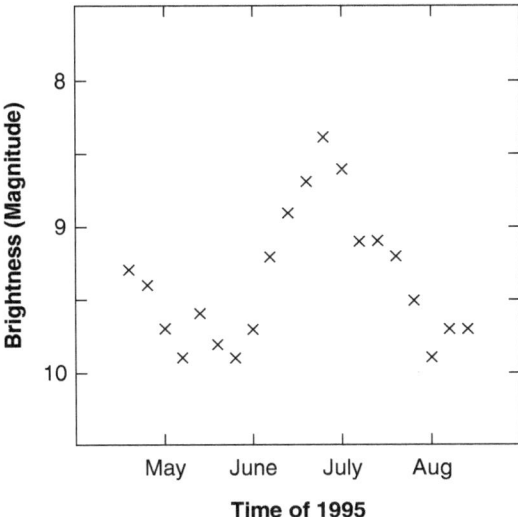

Fig. 1.1 Simple light curve (Image from https://imagine.gsfc.nasa.gov/science/toolbox/timing1.html)

to an astronomer and a tool that reveals much about the system under scrutiny. An example can be seen in Fig. 1.1.

The record of changes in brightness that a light curve provides can help astronomers understand processes at work within the object they are studying and identify specific categories (or classes) of variable events. Thanks to successive generations of astronomers studying variables and drawing light curves based upon observation, astronomers know generally what light curves look like for a class of variable star. In this fashion, when a new light curve of a variable star is plotted, we can compare it to standard light curves in order to identify the type of star under observation.

The gifted British astronomer John Goodricke discovered the variability of both δ Cephei and β Lyrae, while his astronomical companion Edward Piggott discovered η Aquilae in 1783. Working in concert with each other and communicating via letters (Goodricke was deaf and mute) Goodricke and Piggott distinguished two classes of variable star. The first type consisted of objects such as Algol, which exhibited a single sharp change in brightness on a regular basis. In the case of Algol, Piggott and Goodricke correctly surmised that the changes in brightness could be explained by transits of some dimmer object across the star, and they even postulated that it might be caused by a transiting planet. This remarkable achievement was tempered once it was known that Algol has a transiting fainter companion star rather than a planet.

The second type the pair distinguished included variable stars such as δ Cephei, whose brightness changed continuously and whose peak brightnesses were not necessarily identical from period to period. They inferred correctly that these irregularities meant that something had to be happening internally to the star, as a transit would produce a regular light curve with no differences between successive periods. Thus they heralded a new field of astrophysical phenomenon that took almost two centuries to understand. Goodricke's notes can be seen here in Fig. 1.2.

By 1786 at least ten variable stars were known. The astronomer William Herschel also drew attention to variable stars and studied the light curves of δ Cephei, β Lyrae and η Aquilae in 1784. In 1787 he discovered that the

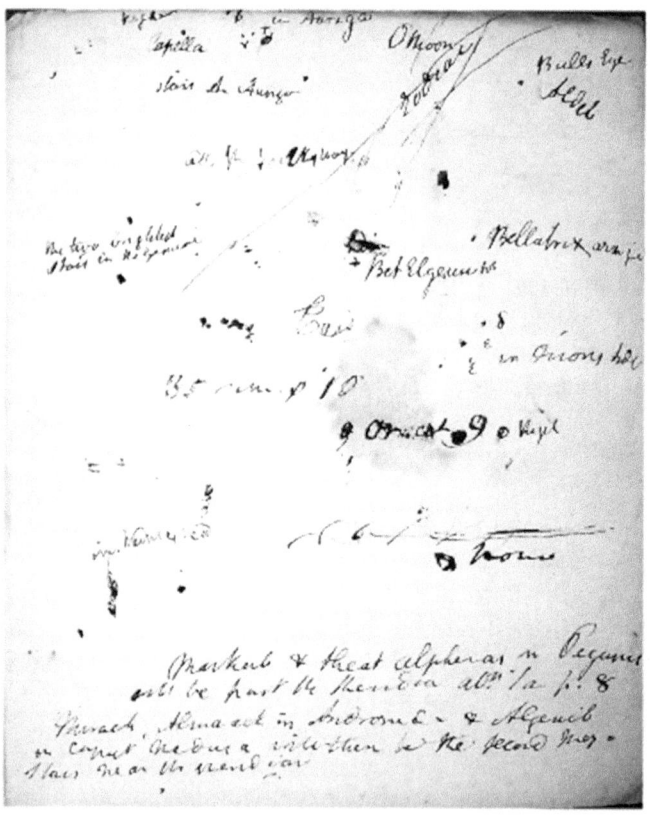

Fig. 1.2 Goodricke's notes of δ Cephei (Image from https://www.researchgate.net/figure/John-Goodricke-1764-1786-Pastel-portrait-by-James-Scouler-now-the-property-of-the_fig1_224861563.)

fainter component of the binary star ι Bootis was variable, and in 1795 he also discovered the irregular variations of the star α Herculis. His son, John Herschel, added to the catalogues of known variable stars with his observations of the southern hemisphere sky from Feldhausen in South Africa in the 1830's.

The following table lists the known variable stars up to the beginning of the 19th century.

Star	Year Discovered	Discoverer
o Ceti (mira)	1596	Fabricius
β Persei	1669	Montanari
χ Cygni	1687	Kirch
R Hydrae	1704	Maraldi
R Delphini	1751	Hencke
R Leonis	1782	Koch
μ Cephei	1782	Herschel
β Lyrae	1784	Goodricke
δ Cephei	1784	Goodricke
η Aquilae	1784	Pigott
R Coronae Borealis	1795	Pigott
R Scuti	1795	Pigott
α Herculis	1795	Herschel

The increase in discoveries in the latter half of the 18th century typifies the way in which scientific observations of the sky were being made in systematic sweeps by people such as William Herschel and Edward Pigott.

It was important that these new types of stars be given some significance so that observers could follow them as often as possible. Therefore, a system was developed by the German astronomer Friedrich Argelander, who gave the first previously unnamed variable in a constellation the letter R, which was the first letter not used by Johannes Bayer in his 1603 *Uranometria*. Today, in any given constellation, the first set of variable stars discovered is designated with letters R through Z. The letters RR through RZ and SS through SZ, and up to ZZ are used for the next variable stars in the constellation. Later discoveries used letters AA through AZ, BB through BZ, and up to QQ through QZ (the letter J is omitted, however). Once those 334 combinations are exhausted, variables are then numbered in order of discovery, starting with the prefixed V335 onwards.

Argelander's seminal star catalogue *Uranometria Nova*, published in 1843, encouraged worldwide interest in the study of variable stars. In fact, Benjamin A. Gould was a pupil of Argelander and became the first American astronomer to receive training in Germany. When he returned to America he encouraged variable star observation and produced his own catalogue of southern variables, called *Uranometria Argentina*, in 1879.

The 19th century saw the application of astronomers such as J. R. Hind in London and the American observers Seth C. Chandler, Edwin F. Sawyer and Paul S. Yendell to variable star work, who all observed from New England.

Chandler produced three catalogues of variable stars, and in 1878 he even produced papers on how to observe variable stars that were circulated among astronomers. In the following years, Chandler continued his variable star work at Harvard College Observatory, although he never became a member of the faculty. Edwin Sawyer was discovered to be a remarkable observer by Chandler, who promoted his interests to produce a catalogue of over 3,400 southern variable stars. The third member of this remarkable trio, Paul Yendell, contributed 140 papers in just 10 years to the journal *Popular Astronomy*. The incredible thing about all three observers was that they were amateurs with no formal training. Chandler was an insurance clerk, Sawyer a bank clerk and Yendell was a shopkeeper! However, being in close association with the director of the observatory at Harvard College, Edward C. Pickering could see the need for a formal society to collate information and disseminate ideas.

Pickering personally made over 6,000 variable star observations and produced catalogues for astronomical use, but by 1910 it was obvious that the sheer number of observations – and observers – required a fresh and organized approach. Eventually it was, once again, an amateur astronomer who took up the responsibility and began to correspond with variable star observers all over the United States. William Tyler Olcott offered his services to Pickering, and an American variable star observer's society began with an article in the November issue of *Popular Astronomy*. The dedication and patience of generations of observers laid the foundations for a society that has become instrumental in amateur and professional variable star work – the American Association of Variable Star Observers, or the AAVSO.

In Great Britain a similar society was inaugurated under the British Astronomical Association (BAA), and it predates the AAVSO. The BAA was founded in 1890, and in the same year a section of variable star observers was set up to encourage observations and produce papers and materials for distribution to other amateurs. The Variable Star Section (VSS) still maintains the aim of collecting and analyzing observations of variable stars. Reports to its members are given via the VSS circulars published four times a year, and there are many articles in the bi-monthly BAA journal.

With the invention of the spectroscope and astronomical photography, the number of known variable stars increased rapidly to the point that now, in the 21st century, the *General Catalogue of Variable Stars* lists more than 46,000 variable stars in the Milky Way alone, as well as 10,000 in other galaxies and over 10,000 'suspected' variables.

Broad Groups of Variable Stars

Variable stars are so different in type and variety that trying to pull them all together into simple headings is a difficult task. However, astronomers noticed that there is a broad distinction that can be used. Whatever happens to cause the variability of any star is due to just two factors: something is happening to the star internally, or something is affecting the star externally. These two reference points then can produce a broad category for variable star discovery once we have a good light curve.

Intrinsic variable stars are stars where the variability in light output is being caused by changes to the physical body of the stars themselves. The following light curve in Fig. 1.3 illustrates the activity of such stars.

As can be seen, the period of variability is over 5.4 days. There a sharp rise to maximum light (maxima), which reveals that the underlying mechanism inflates the star quite quickly to maximum size and luminosity. As the stellar surface cools and relaxes, the star returns to normality in a smooth decline that is not as sharp as the original rise. This shows that radiation is escaping the star in a gradual process almost like a release valve, allowing the surface to return to normal in a controlled fashion before the

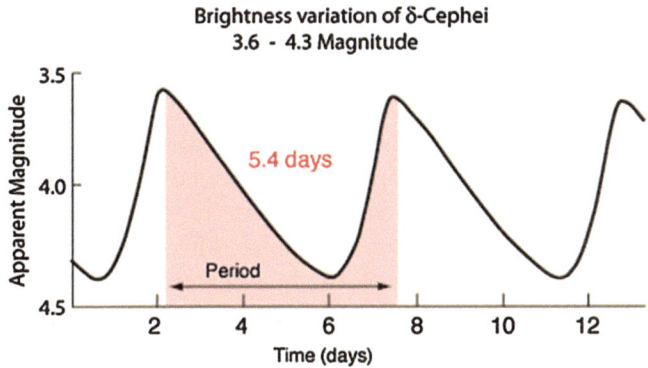

Fig. 1.3 Intrinsic variable light curve (δ Cephei). (Image from http://hyperphysics. phy-astr.gsu.edu/hbase/Astro/cepheid.html.)

whole process starts again. We will deal with the mechanism of expansion and contraction in the chapter on the astrophysics of such objects.

Overall, intrinsic variables follow a general pattern of the above light curve, with subtle or extreme changes dependent on the type. Intrinsic variable stars can be divided into three main subgroups:

Pulsating variables:

Wherein the star's radius alternately expands and contracts as part of its natural evolutionary processes. The stars literally swell in size before declining back to their (almost) original size (see Fig. 1.3).

Eruptive variables:

These are stars that experience physical eruptions on or from their surfaces such as flares or mass ejections.

Cataclysmic or explosive variables:

These are stars that undergo a cataclysmic change in their properties, like novae or supernovae. Although some of these types interact with a companion star they fall within the intrinsic variable bracket, as the mechanism of variability happens to the main star.

The second major group of variable stars is what are called extrinsic variable stars. These are stars where the variability is caused by external properties such as rotation or eclipses with a binary companion. The illustration here in Fig. 1.4 shows the mechanism behind an eclipsing variable star.

This is the light curve of Algol, a typical eclipsing binary and one of the most studied objects in the heavens. As Piggott and Goodricke discovered, the light curve and period of variability are very constant.

When the line is flat, both stars are visible from Earth; then a primary eclipse begins as the fainter companion stars moves in front of the primary and causes the light output to decline rapidly. In some eclipsing variables there will be a flat bottom to the prime eclipse. This shows that the eclipsing star takes some time to move across the line of sight of the primary. Details such as this give us such information as relative size of each star, the orbital period and a host of other factors.

Once the companion moves away from the primary, a sharp increase to normal light is achieved before the secondary star goes into eclipse behind the primary, and a smaller, shallower eclipse of the fainter star is seen before

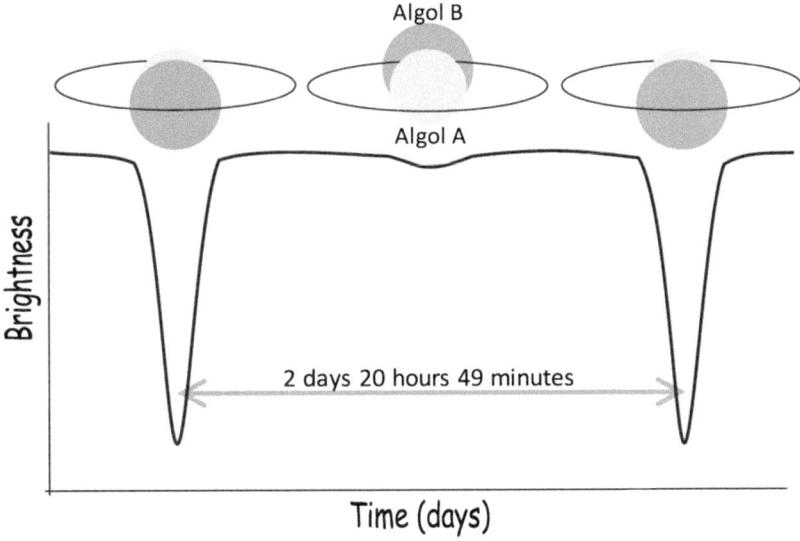

Algol B

Algol A

2 days 20 hours 49 minutes

Brightness

Time (days)

Fig.1.4 Eclipsing variable (Algol) (Image from http://www.adirondackdailyenterprise. com/opinion/columns/2016/12/look-into-medusas-eye-for-the-demon-star/.)

the orbit brings both into view and the light output returns to normal once more.

With extrinsic variable stars there appear to be two main subgroups:

Eclipsing binaries:

These are double stars where, as seen from our vantage point here on Earth, the stars occasionally eclipse one another as they orbit their common centers.

Rotating variables:

These are stars whose variability is caused by phenomena related to their rotation. Rotating variables may be subject to such phenomena as extreme 'sunspots,' which then affect the apparent brightness of the star or in some cases they are stars that have fast rotation speeds, which cause them to become ellipsoidal, or egg shaped!

In both intrinsic and eclipsing variable stars, the subgroups themselves are further divided into specific types of stars that are usually named after their prototypes.

Catalogue Classifications

The common types as seen above are of course subdivided due to type and into subgroups that illustrate the peculiarities of some of the variable star systems. There are additional identification markers that illustrate the wide variety of variable stars and also show that just one sort of variability is not inimical to some systems. Several sub-types show behavior that is typical of several different types, and as a result, these features need to be illustrated for correct classification.

The most important reference source for variable stars is the *General Catalogue of Variable Stars* (GCVS), which contains data for 52,011 individual variable objects discovered and named as variable stars by the year 2015 and located mainly in the Milky Way. From this source is taken the *International Variable Star Index* (VSX) used by the AAVSO, and one can register and scan the VSX at this website: https://www.aavso.org/vsx/. It is instructive to note that variable classes and types are all in uppercase bold letters, so a star such as R Coronae Borealis will be known as RCB, while γ Cassiopeia systems will be GCAS. Some measure of knowledge of constellations is required to tease the names out, but this should be *de rigeur* for amateurs who are going to undertake such work with variables. To learn more about the *General Catalogue of Variable Stars* and to peruse it, then check out this website: http://www.sai.msu.su/gcvs/gcvs/vartype.htm.

Some of the types of variable stars that one will encounter in this book are placed here as a rough and quick guide with some examples of the nomenclature given for certain stars. We shall examine these types in more detail in each chapter on them.

Eruptive variables:

BE, FU, GCAS, I, IA, IB, IN, INA, INB, INT, IT, IN(YY), IS, ISA, ISB, RCB, RS, SDOR, UV (UV Ceti), UVN, WR (Wolf-Rayet types).

It should be noted that the I types here are generally poorly studied and amateurs with the correct photometric equipment may make some valid scientific contributions to their field of study. Many of them (IN to INYY) are commonly known as Orion-type variables, so we are generally looking at young objects.

Pulsating variables:

ACYG, BCEP (β Cephei), BCEPS, BLBOO, CEP, CEP(B), CW, CWA, CWB, DCEP, DCEPS, DSCT, DSCTC, GDOR, L, LB, LC, LPB, M (Mira types), PVTEL, RPHS, RR, RR(B), RRAB, RRC, RV, RVA, RVB, SR, SRA, SRB, SRC, SRD, SRS, SXPHE, ZZ, ZZA, ZZB, ZZO.

The pulsating variables are a fascinating group, as they generally pulsate in regular fashion in a radial expansion. However, some types, such as L to LPB, reveal irregular behavior that is probably due to non-radial pulsation. Sections of the star are moving inward and outward in different modes, and so the star is no longer a spherical object. Some Mira-type variables exhibit such behavior.

Rotating variables:

ACV (A Canes Venaticorum), ACVO, BY, ELL, FKCOM, PSR, R, SXARI.

Rotating variables, as the name suggests, are stars without a uniform surface brightness, although the mechanism for some of their variability remains unclear. Most often the periods are due to the stars being ellipsoid in shape due to fast rotation. Occasionally, the stars vary as large spots or groups of spots are brought into view by the rotation of the star, or there may be some form of thermal or chemical differences in the photosphere or chromosphere, possibly created by a magnetic field. It is thought common to these stars that these intense magnetic fields may not have polar axes in the same plane as the rotational axis.

Cataclysmic variables:

N, NA, NB, NC, NL, NR (recurrent nova), SN, SNI, SNII, UG (U Geminorum types), UGSS, UGSU (SU Ursae Majoris types), UGZ, ZAND (Z Andromedae).

Irregular outbursts characterize these stars, and the group contains not just the supernovae types but the more common U Geminorum types, known as dwarf novae. The UGSS types are SS Cygni stars while the UGZ are very interesting Z Camelopardalis-type stars. After outburst they do not always return to their original luminosity but remain a magnitude or so above their mean.

Eclipsing binary systems:

E, EA, EB, EP, EW, GS, PN, RS (RS Canes Venaticorum), WD (stars with white dwarf components), WR (eclipsing Wolf-Rayet stars), AR, D, DM, DS, DW, K, KE, KW, SD.

Many eclipsing variables are stars that evolve within binary systems and thus fill an area known as the Roche lobe. If they do this as they expand, then materials can tip over the inner Lagrange point and mass transfer begins. The characteristics of such light curves are complicated by rotation around their gravitational centers, the spread of material masking the light output of the stars and the contribution to the light curve from hotspots in accretion discs. Obviously, they are not as simple as the typical Algol system!

Additional methods of identification are used in both catalogues. For example, an upright character (I) between two different types gives the distinction "or" if the classification of the star is uncertain. A typical example of this is ELLIDSCT, where the star may be an ellipsoidal binary system or a δ Scuti-type pulsating variable.

The symbol + means that there are two different variability types in the same star system. An example of this would be ELL+DSCT, where one of the components of an ellipsoidal binary system is again a δ Scuti-type pulsating variable.

The slash (/) symbol indicates a subtype of star. In the case of binary systems, it is used to help describe either the physical properties of the system (E/PN or EA/RS) or the luminosity class of the components (EA/DM). These are some common components of the standard General Catalogue of Variable Stars classification system.

Most star designations are self-evident as typical of their sub-types or are named after the progenitor star of their type. So, as examples, we can discern non-eclipsing RS Canes Venaticorum-type stars as (RS), Wolf-Rayet stars as (WR) and so on. For some cataclysmic variable stars, the subtypes of NL or "nova-like" are designated NL/V, or NL/VY are designated as independent classes since these subtypes do not apply to a class other than NL.

We have to be as careful and as accurate as possible in defining variable subtypes, as many variables show more than one type of variation. In order to distinguish these types, it is easier to identify the primary cause of variability and use that as the basis of classification. For example, S. Doradus-type stars (SDOR) show both eruptions and pulsations, but they have been included in the eruptive group only. The type of definition gives an instant explanation of their chief behavior. Another example can be seen from the R. Coronae Borealis (RCB) stars, which also pulsate. However, their main variable feature is a deep plunge from their normal magnitude to something very dim; then they are included in the eruptive group only. Similarly, the mechanism of variability of the λ Eridani types (LERI) is not well known. It is thought that their variability may be due to either rotation or pulsation, but they are included in the class of rotating variables only.

Now that we have looked at a brief history of their discovery and a basic explanation of variable star types, perhaps it is instructive to ponder exactly what is going on with these stars. Why do they vary and what are the underlying mechanisms that drive such behavior? That will be the subject of our next chapter.

Chapter 2

The Astrophysics of Variable Stars

Stars experience stages of birth, growth, middle and old age and finally death. Astronomers talk of these stages as progressive stellar evolution, although it bears no resemblance to the biological theory of evolution proposed by Darwin. They begin with nebulae and generally end with a nebula, whether it be a planetary nebula or the expanding mass of a supernova explosion. Stars are the only entities in our universe that follow the strict rule of evolution, slow change with time, but they always remain stars, of course, for the majority of their lives.

The remarkable thing about stars is that, while they remain constant for long periods of time, during their birth phase and during their mature years they all undergo variability of some kind. The different forms of variability have given rise to many types of variable class, some of which are tied together due to age and general features of the spectral type, some of which are dependent on their mass and others which are dependent on age. Variable stars can also change their spectral class and color as they vary, while most stars undergo some transformation in color and luminosity as they either settle down to a long life or age and die. Knowing something about the life cycle of stars and their origins will therefore inform us as to the types of variable stars we are observing.

© Springer Nature Switzerland AG 2018   13
M. Griffiths, *Observer's Guide to Variable Stars*, The Patrick Moore
Practical Astronomy Series, https://doi.org/10.1007/978-3-030-00904-5_2

Population Types

A classification that is important to the nature of the variable is the grouping of stars into what has become known as stellar populations. Thankfully, there are only two of these classes. This rule of thumb has arisen due to the discovery that not all stars are equal in their chemical composition. Many stars that have great ages were found to be deficient in many common metals such as sodium, magnesium, etc. This is entirely due to their periods of origin. Older stars are of relatively low mass and are found in the nucleus and halo of the Milky Way Galaxy in addition to being members of the globular clusters.

These stars are known to be at least 10 billion years old or more, and to have formed at a time when our galaxy, indeed most of the universe, was forming. As the greater majority of galaxy building material was hydrogen, then these stars are primarily hydrogen based, whereas stars of later generations, while also predominantly hydrogen, have a large admixture of metals. These second generation stars are to be found only in the discs of galaxies, where nebulae and clouds of dust abound to create more young stars.

Thus astronomers classify old halo and nuclei stars as Population II, and young stars of the galactic disc are called Population I. The classification is easy to remember if you consider their location in any galaxy:

Population I - Stars of the galactic disc and star clusters
Population II - Stars of the galactic halo, galactic nucleus and globular clusters

Variable stars of the Population I group include classical Cepheid variables, the majority of long period variables (LPV) and luminous blue variables (LBV). Variable stars of Population II are RR Lyra-type variables, W Virginis stars and some LPV's. We shall examine these stars in due course, but it may be instructive to understand how stars are classified and know the story of stellar evolution.

Stellar Classification

Most of the public is completely unaware that stars have different colors. Color perception is a personal thing, and just a little training by examining different stars will enable one to identify these colors quite easily. This color difference is extremely important as color is related to the stellar temperature.

The first time stars were grouped according to color was at Harvard University at the end of the 19th century. The eminent astronomer Edward Pickering brought together a fine group of young ladies to do the work of assisting him in his quest to catalogue stars by spectral appearance. This was an unusual step to take, when astronomy was generally considered a gentleman's pursuit. As a result, the group become affectionately known as "Pickering's Harem," but these young ladies proved to be a valuable resource; several of them, such as Henrietta Leavitt, Cecilia Payne-Gaposhkin and Annie Jump Cannon, made discoveries of literally cosmic importance as their work moved into the 20th century.

Recognizing the importance of color as a function of temperature, the Harvard ladies first thought to catalogue the stars by using the English alphabet A, B, C, D, etc. However, the constraints placed upon the colors and spectral characteristics of stars led to the representation that is used today, as many of the original classifications turned out to be false leads or were repetitions of other stars. This classification by color revealed that the hotter the star, the more blue it was; the cooler the star, the redder it was. This reflects the electromagnetic spectrum itself, where the component of visible light that we see between 390 and 700 nm (nanometer or nm = 1 billionth of a meter or 1×10^{-9} m) follows the grouping of colors familiar to us from the rainbow in Fig. 2.1. The shorter (violet) end of the spectrum is at 400 nm and the longer wavelengths (red) are close to 700 nm.

The grouping that the Harvard computers eventually settled on as representative of all stellar spectra, using the strength of the hydrogen Balmer lines and Wein's law as a guide is simply this: O B A F G K M, as seen in the illustration here. These are taken singly below:

Type 0 stars are blue-white and extremely hot, typically around 25,000 K and higher. These quite massive stars are very luminous and are the most

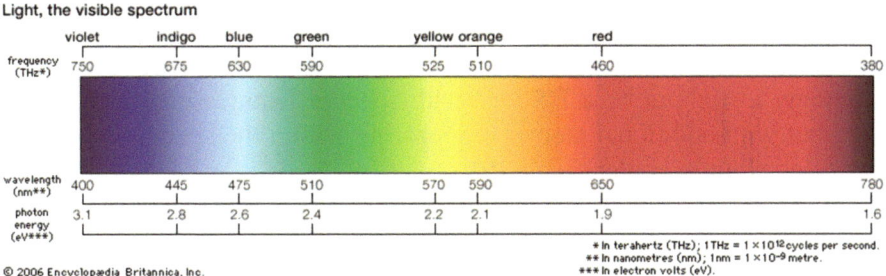

Light, the visible spectrum

© 2006 Encyclopædia Britannica, Inc.

* In terahertz (THz); 1THz = 1 × 10¹² cycles per second.
** In nanometres (nm); 1nm = 1 × 10⁻⁹ metre.
*** In electron volts (eV).

Fig. 2.1 The visible spectrum

short-lived of stars. Their spectra show lines from ionized helium, nitrogen and oxygen. A typical example of this class is Iota Orionis.

Type B stars are blue-white and very hot, with temperatures of around 20,000 K. Stars of this type are generally massive and quite luminous. Their spectra display strong helium lines. Rigel, Spica and Regulus are good examples of this class. Compare Rigel and Regulus as an exercise.

Type A stars are white, with temperatures of around 10,000 K. Their luminosities are usually about 50 to 100 times that of the Sun. At these temperatures no helium lines are present, but strong hydrogen lines appear in the spectra. Sirius, brightest star in the night sky, is of this class as are Vega and Altair, part of the summer triangle.

Type F stars are yellow or yellow-white, with temperatures of about 7,000 K. Their spectra show weaker hydrogen lines, but strong calcium lines. In some of the later classes the spectra of metals begin to appear. The winter star Procyon in Canis Minor is of this class.

Type G stars are yellow, with temperatures of about 6,000 K. Their spectra show weaker hydrogen lines, but stronger lines of many metals. The sun is a typical G-type star.

Type K stars are orange, with temperatures of about 4,000 to 4,700 K. They have faint hydrogen lines, strong metal lines and hydrocarbon bands in their spectra. Aldebaran in Taurus and Arcturus in Bootes are good examples.

Type M stars are red, with temperatures of about 2,500 to 3,000 K. They have many strong metallic lines and wide titanium oxide bands, with other exotic compounds in their spectra. Betelguese and Antares are spectacular and easily visible examples of this class.

There are three other classes, namely R, N and S, now referred to as C-type stars, but these refer to fairly rare types of stars. A useful way of remembering the classification is by using the phrase: **O**h **B**e **A** **F**ine **G**irl **K**iss **M**e **R**ight **N**ow **S**mack! The groups can be seen in the H-R diagram below. Each of the classes are further divided into the numbers between 0–9 so that a better approximation can be made of the spectral type. Under this taxonomy a star can be a B5 type with spectral features of the typical B class but with additional elements appearing within the spectra due to the star being slightly cooler than a B0.

Due to this color and temperature relationship, it is possible to take the peak visible output of a star in nm and work out the peak wavelength of light – or its dominant color by using a mathematical relationship first postulated by the German physicist Wilhelm Wein in 1893. This is known as Wein's law. Simply put, Wien's law is the relationship between the peak

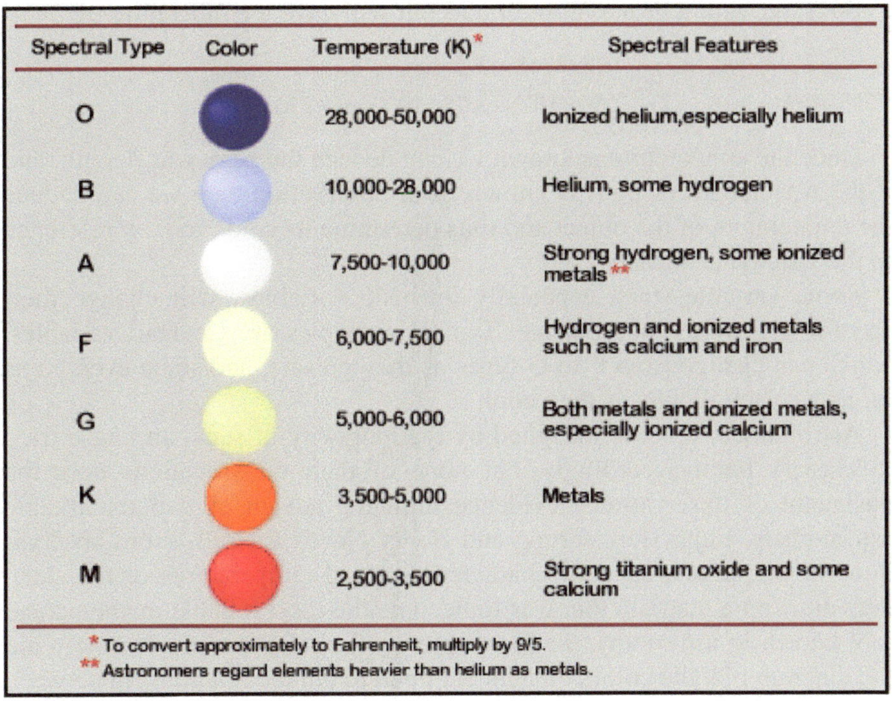

Spectral Type	Color	Temperature (K)[*]	Spectral Features
O		28,000-50,000	Ionized helium, especially helium
B		10,000-28,000	Helium, some hydrogen
A		7,500-10,000	Strong hydrogen, some ionized metals [**]
F		6,000-7,500	Hydrogen and ionized metals such as calcium and iron
G		5,000-6,000	Both metals and ionized metals, especially ionized calcium
K		3,500-5,000	Metals
M		2,500-3,500	Strong titanium oxide and some calcium

[*] To convert approximately to Fahrenheit, multiply by 9/5.
[**] Astronomers regard elements heavier than helium as metals.

Fig. 2.2 Spectral classes

wavelength of electromagnetic radiation from an object and its temperature. It can be expressed as:

$$\lambda = b / T$$

where λ is the wavelength, b is Wien's displacement constant of 2.897768×10^{-3} m K^{-1} and T is the temperature in Kelvin (K).

If a star has a temperature of 6,000 K, what is its peak output? (For quick results ignore the decimal point and the power):

$$2897768 \div 6000 \ \text{K} = 482 \ \text{nm}$$
$$\text{Or}: 2.89 \times 10^{-3} \div 6000 \ \text{K} = 4.816 \times 10^{-7} \ \text{m} \ \text{or} \ 481.6 \ \text{nm}$$

Conversely if you know the peak output, one can work out the temperature by commuting

$$T = b / \lambda$$

So a star with a peak output of 375 nm will have a temperature of:

$$2897768 \div 375 \text{ nm} = 7727 \text{ K}$$
$$\text{Or}: 2.89 \times 10^{-3} \div 3.75 \times 10^{-7} \text{ m} = 7706 \text{ K}$$

Once the temperature is known we can deduce the peak wavelength, and if the wavelength of light is known from observation then we can deduce the temperature of the object and thus determine its color, too, by reference to the spectral sequence.

Some variable stars, especially intrinsic variables, will change their spectra slightly as they pulsate. Typical examples are Cepheid variables, which can change from F to G types as they pulsate, and some even go as far as K spectral class as they cool.

Astronomers become intrigued by the longevity of stars, and have tried to classify them accordingly, but none of their classifications bore the onslaught of observational evidence until the advent of two remarkable astronomers, Enjar Hertzsprung and Henry Norris Russell. From observations of the colors, spectral characteristics and temperatures of the stars, they drew up a diagram that was to become the most reliable mathematical tool known in astronomy, the Hertzsprung -Russell diagram. Not only did the diagram plot the colors of stars but pointed out the relationship between color, temperature, mass, luminosity and age. The diagram is a very powerful tool indeed; only a few constraints have to be well established about any particular star, and its consequent life story can by prophesied in detail.

You will notice that luminosity increases upward on the left hand Y-axis of the diagram and the color and temperature are plotted on the X-axis. On the right hand Y-axis can be plotted absolute magnitude so that a relationship between luminosity and peak absolute magnitude can be judged. This scale is not included here merely for simplicity. One immediately notices that the stars form an "S"-shaped curve from top left to bottom right – a curve that is known as the main sequence, where stars are shining in equilibrium and converting hydrogen into helium in their cores. Stars spend the great majority of their lives on the main sequence.

Above and to the right of the main sequence is the giant branch, where stars that have exhausted their hydrogen fuel lie. Eventually they will convert helium into carbon in their cores and become "horizontal branch stars" trying to get back to the main sequence. However, the time required to cross back to the main sequence is longer than the helium burning stage, and so they then become a supergiant star if they have masses approximating to that of 1 to 5 solar masses (M_o). Although it is not depicted here, as such stars cross the horizontal branch they encounter an area on the H-R diagram known as the instability strip. Here the stars begin to pulsate, becoming

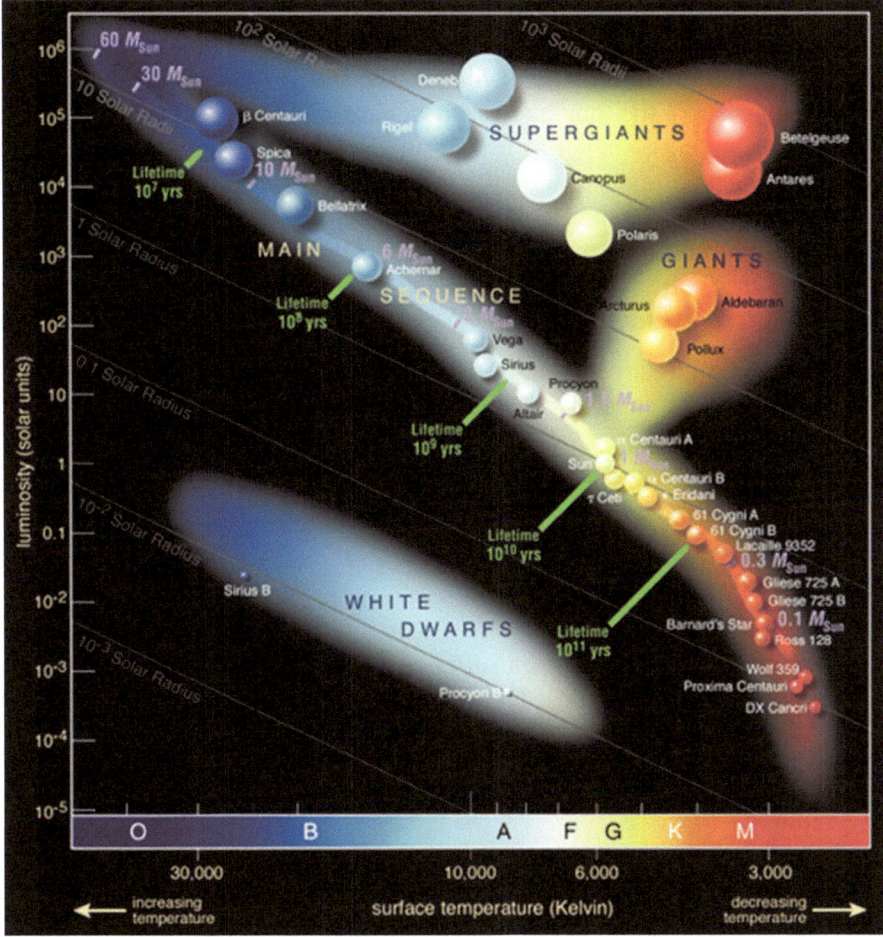

Fig. 2.3 HR Diagram

intrinsic variable stars with periods ranging from a few hours to a few days depending on their mass. This is where the Cepheid variable stars lie, which we shall examine later.

Stars in the above mass range will become supergiant stars (known as Asymptotic Giant Branch stars) rich up in the top right hand corner of the HR diagram. These stars will eventually lose mass and form planetary nebulae before contributing their burned out cores to the ranks of the "White Dwarfs" seen at the lower left side of the diagram.

Stars in the large mass ranges on the main sequence in the higher left hand side of the diagram follow similar pathways, though it is more likely that due to their high mass they will become supernova – enormous explosions that destroy the star and leave behind either a highly compressed remnant of the core known as a neutron star or if massive enough, could become a black hole.

These groupings on the HR Diagram can be further subdivided into notations that indicate the size and luminosity of the star and are known as the MKK classes after William Morgan and Phillip Keenan at Yerkes Observatory. They are:

Ia - Most luminous supergiants
Ib - Less luminous supergiants
II - Bright giants
III - Normal giants
IV - Subgiants
V - Main sequence
VI - Subdwarfs
VII – White dwarfs

In addition to these classes, a suffix can be attached to a star to describe any peculiar states or rarities. The suffixes commonly used are:

d - Dwarf stars
G - Giant stars
D - White dwarfs
E - Emission lines
p - Peculiar stars

Therefore, one look at the star's position on the H-R diagram can tell one a lot about the star; its temperature, spectral type MKK class and luminosity, give some judgment on its radius and also define what the end product of its evolution will be. The H-R diagram is a very powerful astrophysical tool. What is more, it is relatively easy to use, explore and memorize.

From a visual perspective, it is worth keeping in mind that star colors are attenuated by several factors, such as seeing in the atmosphere, turbulence of the air, the elevation above the horizon, etc. Some observers are disappointed that red stars do not appear really so red either! This is due to the fact that, as seen above, spectral typing and thus apparent color is based on temperature. The hotter the star, the whiter or bluer it will appear. Stars in the spectral range of O to A usually have temperatures that descend respectively from 25,000 K to around 10,000 K. Stars of F and G type have temperatures between 7,000 K to 5,500 K.

Using Wein's law, we can see that cool stars actually have their peak output in infrared; as humans do not perceive that part of the spectrum, a red variable star may therefore not appear very red to the naked eye. Conversely, blue stars have a peak output in the ultraviolet and should appear purple in color; but again, the human eye's sensitivity does not extend into that area of the spectrum, so they appear blue-white to us.

The peak radiation output thus accounts for the disappointment of many observers hoping for strikingly "blue" or "red" stars.

Red stars are a wonderful introduction to attempting to discern color and hue, and it is worth noting that such stars vary in their "redness." Some are more orange, whereas others can be an intense red. For example, Mu Cephei was called the "Garnet Star" by William Herschel, who thought it to be a deep ruby color, whereas J. R. Hind, a 19th century observer, described R Leporis as "an illuminated drop of blood." Such intensely red stars were formerly classed as R, N or S on Canon's stellar classification scale and are now known to be stars undergoing variation. Modern astronomy recognizes that these stars contain a high quantity of carbon in their atmospheres and have now been re-classified as "C" stars, the C standing for the chemical element carbon. Nevertheless, intensely red stars are a visual rarity.

To understand the variability of stars at the beginning and end of life, it is necessary to briefly explore the life cycle of stars and touch upon their variable properties as we go along.

Stellar Evolution

Exactly like life on Earth, stars experience stages of birth, growth, middle and old age and finally death. Astronomers talk of these stages as progressive stellar evolution, although it bears no resemblance to the biological theory of evolution proposed by Darwin. Stars are the only entities in our universe that follow the strict rule of evolution, slow change with time, but they always remain stars, of course.

The birth of stars is shrouded, not so much in mystery but by the clouds that they are born from. These clouds are visible as dark nebulae along the spine of the Milky Way and are the nurseries of stars yet to come. Stellar birth is a process hidden in darkness, and it is only with advances in infrared, radio and short wavelength astronomy that we have been able to see increasing detail in such clouds and add to our knowledge of star birth.

As we have discovered above, a most useful tool enabling astronomers to make predictions of stellar evolution is the Hertzsprung-Russell diagram. From an observational point of view, just a few parameters need to be met to place any star on the HR diagram and so enable an examination of the timelines of birth, lives and deaths, the lives of stars from cradle to grave so to speak. Features such as color relate to temperature, and so both these are included in the diagram, as are the luminosity in comparison to the Sun and the absolute magnitudes of any stars. These can be obtained by distance modulus calculations so that the real luminosity of stars can be gauged and the placement of a body on the diagram made as accurately as possible. It is possible to use the diagram across a wide spread of stellar ages; the initial sequence of stellar birth does not appear on the diagram but can be explored below.

The Interstellar Medium

It is not difficult to get the impression that the space between the stars contains absolutely nothing at all, but in fact this could not be further from the truth. The spaces between the stars contain the gas and dust of the interstellar medium (ISM), the future materials of suns and planets. Though this material is thinly spread it is abundant enough due to the vast size of space to form an appreciable mass along our line of sight and extinguish stars in the far distance by absorption of starlight.

However, this material is not always evenly spread across space, and local condensations into vast molecular clouds or knotty dark globules can be seen in certain areas of the sky. Generally speaking, the ISM is quite vapid, with less than one hundred particles per cubic meter. A dense part of the ISM could even reach 10^{17} particles per meter, which in comparison sounds enormous until one considers that in a single average human breath, there are 10^{24} particles! Other considerations are the temperature differences in the ISM, which can range from just 10 K to 1,000,000 K depending on the material under examination. This substance is basically hydrogen and helium gas with trace amounts of carbon, oxygen and nitrogen as well as other elements of the periodic table in very low abundances. The following figure shows the distribution of the ISM in the Milky Way Galaxy; note how the denser areas of the ISM are concentrated along the disk of the galaxy itself.

The ISM is mostly accounted for by matter that can be termed the intercloud medium, a phrase that highlights the difference between the ISM and denser areas such as molecular clouds. The intercloud medium is mostly constructed from hot gasses that are the result of stellar winds and coronal ejections from stars. Included in the intercloud medium are the ejected materials from supernovae and planetary nebulae, and as this material is very sparse, it is possible to observe distant objects through it. Following these low density areas are those of higher density in which potential collapse into new stars can occur or has occurred. These are known as diffuse clouds and dense clouds, while others in this group are known as emission or HII regions; stars have already been born here and are illuminating the nebulae we see.

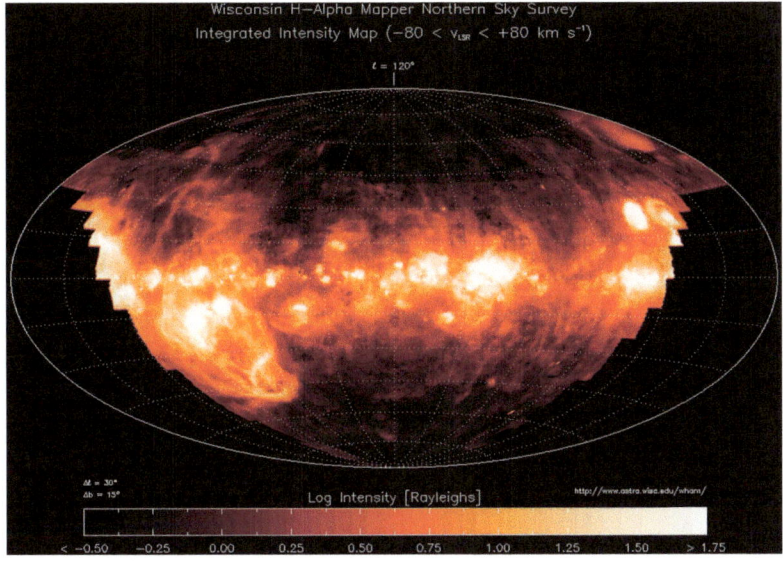

Fig. 2.4 The ISM (Image from https://en.wikipedia.org/wiki/Interstellar_medium.)

The following table illustrates the densities and abundances of the different materials that collectively make up the ISM and how astronomers observe them.

Material	Volume	Temperature	Density (M³)	Hydrogen	Observation
Molecular clouds	< 1%	10–20 K	10^2–10^6	Molecular	Radio and IR
Cold neutral medium	1–5%	50–100 K	20–50	Neutral atomic	21 cm Radio
Warm neutral medium	10–20%	6000–10^4 K	0.2–0.5	Neutral atomic	21 cm Radio
Warm ionized medium	20–50%	8000 K	0.2–0.5	Ionized	Hα emission
HII regions	< 1%	8000 K	10^2–10^4	Ionized	Hα emission
Coronal gas or the hot ionized medium	30–70%	10^6–10^7 K	10^{-4}–10^{-2}	Ionized (and ionized metals)	X-rays

Much of the early work on the ISM was developed by Johannes Hartmann around 1904 after he observed the weak sodium lines emanating from delta Orionis apparently being absorbed and postulated a thin, ether like material lying along our line of sight to the star. His study and those of later astronomers, notably Edward Pickering, using absorption line spectroscopy, put the ISM firmly on the astronomical map; indeed it was Pickering who identified the ISM as a gas rather than the tentative ether idea that was common at the time. Later techniques in IR and radio astronomy have added greatly to our understanding of the ISM and its role in star formation and nebulae production.

However, how does the interstellar medium collapse to form stars in the first place? Why is there still so much of the ISM around if gravitational collapse pulls the materials together?

Cloud Collapse and the Jeans Mass

If it were merely a question of gravity then the ISM might not exist in great quantities today, and the galaxy would be filled with stars in profusion. However, as we can see from the above table, the ISM has a low temperature that maintains equilibrium between thermal pressure pushing the ISM apart and gravity pulling it all together. It is only when gravity becomes ascendant over this thermal pressure that the ISM can collapse into denser areas and begin to form nebulae and stars.

The British physicist James Jeans approached this problem in the early 20th century and determined that there has to be a range of masses for a cloud to collapse and that this mass was determined by the amount of materials present and the volume of space they occupied. Together these quantities are known as the Jeans Mass and the Jeans Length; collectively we can call the conditions that determine star formation in the first instance to be the Jeans Criteria. It is as simple as how much matter is present in a particular volume, and does this material have a low enough temperature to enable gravity to overcome the thermal pressure.

One can see from the above table that dense cold clouds have the best chance of forming stars. It seems obvious that molecular clouds and cold neutral clouds have the best chance of making nebulae that will create new stars. Indeed, this is what E. E. Barnard pointed out in the 1920's during his survey of dark nebulae that the density of such clouds must be relatively high and therefore could collapse. Observations of such areas as dark nebulae and emission nebulae reveal additional small dense clouds of materials of less than 1 pc in diameter that are now called Bok globules after Bart

Bok, the astronomer who studied them and wrote a seminal paper on their physical characteristics in 1947. It was Bok that connected the collapse of the ISM to dense cool clouds, and subsequent observations in infrared in the 1990s confirmed that there are new stars inside Bok globules.

Let us return to James Jeans. His criteria stated that there has to be a specific length and density of particles to begin collapse and that the cloud temperature has to be relatively low. The Jeans length is therefore:

$$R_j = \left(kT / Gm^3 n\right)^{1/2}$$

where k is the Boltzmann constant – 1.38×10^{-23} JK^{-1}
T is the temperature in Kelvin
G is the gravitational constant – 6.67×10^{-11} Nm^2 kg^{-2}
M is the mass of the hydrogen atom – 1.67×10^{-27} kg
n is the number of particles or the number density.

If we take a cloud with a temperature of 50 K and a density of 10^{12} hydrogen atoms per cubic meter, what is the length?

$$R_j = \frac{1.38 \times 10^{-23} \text{ JK}^{-1} \times 50 \text{ K}^{1/2}}{\left(6.67 \times 10^{-11} \text{ Nm}^2 \text{ kg}^{-2}\right) \times \left(1.67 \times 10^{-27} \text{ kg}\right)^2 \times \left(10^{12}\right)}$$

Then $R_j = 6.1 \times 10^{15}$ meters or about 0.19 pc – so less than one light year across.

The Jeans mass can then be found by multiplying the density by the volume, where the volume of a sphere is 4/3 π r^3

$$\text{So: } 4.18 \times 1.67 \times 10^{-27} \text{ kg} \times 10^{12} \times \left(6.1 \times 10^{15}\right)^3 = 1.5 \times 10^{33} \text{ kg}$$

As the mass of the Sun is 2×10^{30} kg then:

$$1.5 \times 10^{33} \text{ kg} = 750 \text{ solar masses}$$

So the Jeans mass for this cloud of a size of less than one light year (0.2 pc) is 750 times the mass of the Sun. The cloud at this point would overcome the thermal pressure of the atoms within it and collapse gravitationally to produce stars.

Generally speaking there are additional factors to consider, such as a "push" from density waves, supernova shock waves or other mechanisms that also impact on the cloud and give it that initial pressure to collapse. As stated previously, the ISM remains relatively stable, and only an external factor such as a shock wave gives the clouds the impetus to condense and collapse.

Once this collapse is underway, the temperature in the cloud rises, but the radiation released by the cloud is generally insufficient to halt the collapse, and gravity wins out to produce the fragments that become future stars. This fragmentation process is important, as it is not known how many stars an individual cloud will produce despite the Jeans mass calculation above giving a general figure simply for the total cloud mass. Under gravitational collapse, small eddies and vortexes produce the conditions for fragmentation of the cloud into smaller globules that then become protostars of varying mass.

From Protostars to the Main Sequence

Once clouds begin to contract under gravity the dust and gas becomes opaque, and the conditions for star formation rapidly gain precedence. These dark clouds or Barnard objects can be seen by amateurs, and several examples can be found later in this book. The mass of such zones varies from 1,000 to 10,000 solar masses while the smaller, fragmented Bok globules have a mass range between 1 and 1,000 solar masses. Their small size reveals that they are collapsing objects while Barnard objects generally remain large molecular clouds in equilibrium, though, of course, the conditions for collapse could happen at any time within them. The connection between Barnard's clouds and Bok globules is now well established by astronomers.

Protostars then are the result of cloud fragmentation into individual units in masses that are larger than 0.1% of the Sun's mass. The internal temperatures of Bok globules remains low and has been measured by radio astronomers to be as low as 10 K, so thermal pressures are not a large consideration during the fragmentation process as gravity wins out once the Jeans mass is established and collapse ensues.

The cloud or globule then becomes convective as the temperature rises and results in an ongoing fight between the warming cloud pushing the material apart and gravity pulling it all together. The convective motion within the cloud gives rise to areas of greater density under the infall of dust and gas. As the density of areas within the cloud rises, contraction continues at a rate faster than the convection and radiation pressure can withstand, and the cloud fragment becomes denser and denser until it becomes metastable and radiates away in infrared. At this point the fragment becomes a protostar.

This process is relatively random and takes place deep within the cloud. The resultant masses of individual protostars is not well known and could vary over the initial lifetime of the protostar as it either gains mass due to continued gravitational infall of material or lose mass as the radiation pressure from the protostar halts the infall of gas and the protostar stops growing. In some cases it could even lose mass due to radiation pressure.

Protostars radiate as they heat up and can be seen within globules as IR and radio emitters. Such stars are not stars in the sense of highly luminous objects we see in the night sky, as they have yet to achieve nuclear reactions in their core. Nevertheless, a protostar of 1 solar mass could be as luminous as 100 times that of the Sun in infrared, and this radiation could have important effects such as stopping further accretion of material onto the protostar. Gravity continues to contract the protostar until the internal temperature and pressures become high enough to initiate nuclear reactions, and at this point the star begins to move toward the main sequence of the H-R diagram and take up station as a real star. Such protostars can be imagined with the aid of an artist's impression, such as the figure shown here:

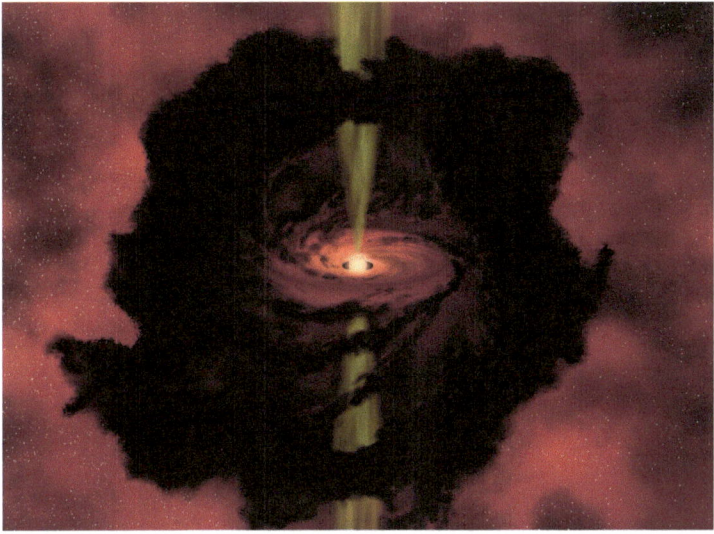

Fig. 2.5 Protostar (Image from https://lco.global/spacebook/protostar/.)

As protostars settle toward the main sequence they go through states of variation typical of the early stars once known as the Orion type variables but now separated into their progenitor types, the FU Orionis-, YY Orionis- and T Tauri-type stars. All three exhibit similar patterns of behavior and are known to be low mass stars leaving the protostar stage. It is worth looking a little closer at the YY Orionis and T Tauri stars to gain an understanding of their variability. In fact, several such early-type stars can be studied by amateurs or followed on an ad hoc basis by casual observers.

Early Variability – The YY Orionis and T Tauri Stars

As protostars settle toward the main sequence, dependent on their mass, they go through a state of variability known either as the YY Orionis phase if they are low mass (<50% of the Sun's mass) or the T Tauri phase after the star T Tauri. The difference between the types is typified by the fact that T Tauri is interacting relatively violently with its surroundings and spraying material away from itself, whereas the YY Orionis stars show profiles of infalling materials, suggesting that they are protostars still in the formation stage. Both stages are tied together as YY Orionis stars will exhibit T Tauri behavior at a later stage. Both stars exhibit magnitude changes that vary between 1 and 2 or even 5 and 6 magnitudes dependent upon type. Therefore, T Tauri variables represent a typical example of these early stars, as long as they fall within a relatively low mass range (between .05 and 3 times the mass of the Sun).

The star throws material off along a polar or magnetic axis and further infall of materials is prevented by radiation pressure. The star loses some mass as material is ejected along these axes. During this phase these stars can be seen deep within dense clouds as Herbig-Haro objects; infrared sources, which are a kind of reflection nebulae where the ejected materials are impacting the surrounding cloud, cause the shocked cloud to glow.

T Tauri itself is surrounded by a variable nebula, which was discovered by J. R. Hind in the 19th century. Such stars are approaching what is known as the zero age main sequence (ZAMS) and will settle into a series of nuclear reactions that will last for hundreds of millions if not billions of years. The ejected materials from such stars follow a similar pattern to the protostar in Fig. 2.5, and the T Tauri ejection can be seen in Fig. 2.6.

Fig. 2.6 T Tauri star (Image from https://en.wikipedia.org/wiki/T_Tauri_star.)

T Tauri stars are essentially variables that show both periodic and random fluctuations in their luminosity. As they are newly formed bodies less than 10 million years old their central temperatures are too low for nuclear fusion to have started. Indeed, for up to another 100 million years, the emitted radiation will come entirely from the gravitational energy released as the star contracts under its own self-gravity. T Tauri stars therefore represent an intermediate stage between real protostars (e.g., YY Orionis stars) and low-mass main sequence (hydrogen burning) stars like the Sun.

The nearest T Tauri stars to Earth can be found in the Taurus and ρ Ophiuchi molecular complexes, both of which are approximately 400 light years away. The prototypical T Tauri itself is part of a close binary system with a smaller, fainter companion. Most T Tauri stars indicate complex interactions with their surrounding environment with hints of stellar winds and huge confined jets imaged in many sources.

Both the winds and jets of T Tauri stars are thought to be powered by material falling onto the central star via a protoplanetary disk that has been observed to surround many of them. Interestingly, in all probability planets will also form from this disk, and some may survive to form a planetary system surrounding the newborn star.

Studies of T Tauri stars suggest a number of sources for their variability, and the timescales of such suggest two mechanisms. Random variations and fluctuations with timescales of a few minutes to some years may be caused by instabilities in the accretion disk, flares on the stellar surface or simply by obscuration from the surrounding dust and gas clouds. The regular variability of T Tauri stars with timescales typically of a few days are almost certainly associated with huge sunspots on the stellar surface that pass into and out of view as the star rotates.

The variations of T Tauri stars provide a rich source of information about the various components of these newly formed star systems. In addition, as a phase of stellar evolution through which our Sun and Solar System experienced 5 billion years ago, the study of T Tauri stars allows a glimpse into the past of both our central star and its planetary system.

Such early type stars eventually settle down and take up position on the H-R diagram conversant with their luminosity, color and temperature, all of which are ultimately dependent on their initial mass. Once settled on the main curve of the diagram, the main sequence, their variability stops (in general terms unless the stars are extrinsic variables) as the orderly process of converting hydrogen to helium enables them to become stable. There are a few things that we can learn from an examination of stars on the main sequence that can give us pointers to the types of intrinsic variables that some stars may become once this stage is passed.

On the Main Sequence

Main sequence stars can come in all shapes and sizes, from giants to dwarfs. However, the one thing that separates them all is their mass, and this feature is the dominant consideration throughout the main sequence as the stellar mass directs the star along its lifetime, whether it be short lived or long lived. Additionally, main sequence stars are characterized by the source of their energy. They are all undergoing fusion of hydrogen into helium within their cores. The rate at which they do this and the amount of fuel available depends upon the mass of the star. Stars on the main sequence also appear to be unchanging for long periods of time. They are extremely stable.

The secret of this stability is known as hydrostatic equilibrium. The inward acting force of gravity is balanced by outward acting forces of gas pressure and radiation pressure. The more massive the star, the greater its gravitational pull inwards. This in turn compresses the gas more. Compressing a gas exerts an outward pressure as the gas resists the compression. In stars this gas pressure alone is not sufficient to withstand the gravitational collapse. However, once the core temperature has reached about 10 million K, fusion of hydrogen occurs, releasing energy in the form of radiation. This exerts an outward radiation pressure, and this combined with the gas pressure balances the inward pull of gravity, preventing further collapse. The star is stable and remains so as long as the equilibrium is maintained.

Stellar mass and dimension requires an accurate yardstick to measure things against. Fortunately, the star we know the best, our Sun, becomes this yardstick, as we know its mass and dimensions very accurately. Mass is the key factor in determining the lifespan of a main sequence star, its size and its luminosity. Therefore, stars are measured in multiples of the solar mass, or percentages of the solar mass, if lower in constitution than our Sun. The mass of a star is the key to stellar longevity, the higher the mass, the shorter the lifespan, while the converse is true for those of smaller mass. The luminosity of stars is a product of their mass to the 3.5 power. Expressed as:

$$L = M^{3.5}$$

This is usually expressed in units of solar mass and solar luminosity as a comparison:

$$L_o = M_o^{\,3.5}$$

If you have a star that is 10 times the solar mass, its luminosity will be $10^{3.5} = 3162\ L_o.$ So a 10 solar mass star will be over 3,000 times more luminous. The equation also works the other way around, too; if you know the luminosity, you can work out the mass by taking the root of the luminosity:

$$M_o = L_o^{\,-3.5}$$

So a star with a luminosity of 20,000 times that of the Sun will have a mass of:

$$M_o = L_o\,20,000^{-3.5} = 16.9 M_o$$

If a star is 12 times the mass of the Sun, then what is its luminosity?

$$L = M^{3.5}$$
$$12^{3.5} = 5985$$

The star is 5,985 times more luminous than the Sun. Conversely if a star is 55,000 times more luminous. What is its mass?

$$55,000^{-3.5} = 22$$

22 times the mass of the Sun

At this point you can also use this knowledge to discover how long stars will stay on the main sequence converting hydrogen into helium. Remember that their main sequence lifetime is highly dependent on their mass. The larger the mass, the greater the core pressure and, subsequently, the greater the nuclear reaction rate. Main sequence lifetimes can be determined with the following expression:

$$T_{ms} = 10^{10} \text{ years} (M / M_o)^{-2.5}$$

where T_{ms} is the main sequence lifetime, 10^{10} years is the lifetime of the Sun and M_o is the Sun's mass. Using this equation we can calculate the main sequence lifetime of stars as a function of the Sun's mass.

If a star is 10 times the mass of the Sun, what is its lifetime in years?

$$(10 \div 1)^{-2.5} = 0.00316$$
$$10^{10} \times 0.00316 = 3.16 \times 10^7 \text{ years}$$

A 10 solar mass star will stay on the main sequence for 31.6 million years.

What if a star is .5 of a solar mass:

$$(0.5 \div 1)^{-2.5} = 5.65$$
$$10^{10} \times 5.65 = 5.65 \times 10^{10} \text{ years}$$

Or 56.5 billon years!!

Clearly, smaller mass stars will live longer while higher mass ones go out in a blaze of glory.

The dominant hydrogen to helium conversion cycle in stars is also a function of their mass. In lower mass stars the process is known as the proton-proton (PP) chain and is dependent on the core temperature being approximately within a 14 to 18 million Kelvin (K) range (in which three types of PP chain become featured – PP-I, PP-II and PP-III – dependent on increasing core temperature). In stars of moderate to high mass (5 or more solar masses) the higher temperatures prevent the hydrogen nuclei sticking together so a chain of catalytic processes known as the carbon-nitrogen-oxygen chain, or CNO cycle, does the same job of converting hydrogen to helium.

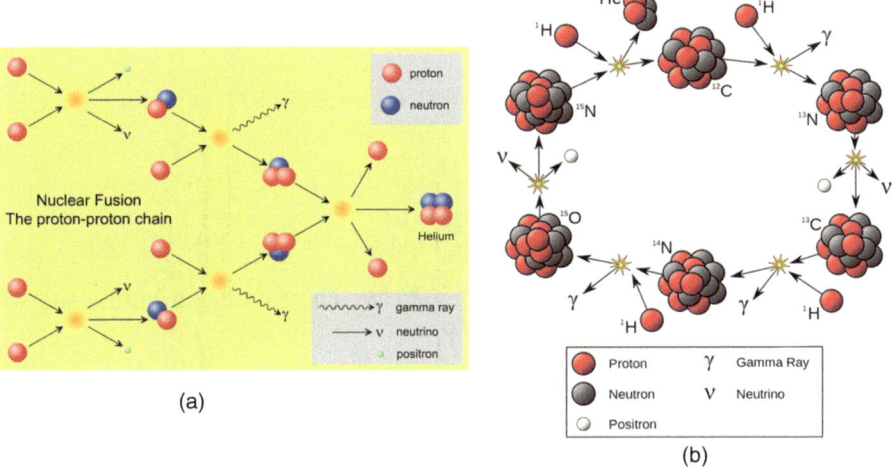

Fig. 2.7 Illustrations of the PP chain and CNO cycle

The stability of these processes cannot be underestimated. Due to these reactions stars will live for millions of even billions of years. It is when the hydrogen fuel eventually runs out that stars undergo a decrease in their stability and states of variability come into play as the stars rearrange their internal processes.

However, it is important to realize, especially in the case of variable stars, that their luminosities are related to their temperature and surface area. Any increase in surface area generally implies a reduction in temperature, so the two are inextricably linked. From this we can determine a number of interesting parameters, such as total luminosity, surface area and temperature, once we know a few things about the star. The luminosity of stars can be worked out as a function of their area and temperature and the Stefan Boltzmann constant. However, the temperature of main sequence stars has to be raised to the 4th power, and the equation is:

$$L = 4\pi r^2 \sigma T^4$$

where σ is the Stefan Boltzmann constant with a value of 5.67×10^{-8} W M^{-2} K^{-4}.

What is the luminosity of a star with R = 500,000 km and temp of 5,700 K?

$$4\pi r^2 = 12.5 \times (500,000\,000 \text{ m})^2 = 3.13 \times 10^{18} \text{ m}$$
$$\sigma = 5.67 \times 10^{-8} \text{ W}$$
$$T^4 = 5,700^4 = 1.05 \times 10^{15} \text{ K}$$

So: 3.13×10^{18} m $\times 5.67 \times 10^{-8}$ W m $\times 1.05 \times 10^{15}$ K $= 1.8 \times 10^{26}$ W.

The star has a luminosity of 1.8×10^{26} W. The Sun has a luminosity of 3.8×10^{26} W, so:

$$1.8 \times 10^{26} \text{ W} \div 3.8 \times 10^{26} \text{ W} = 0.47 \text{ L}_\text{o}$$

The above star is half as bright as the Sun.

What about a star with a radius of 150×10^6 km and a temperature of 3,000 K?

$$150 \times 10^9 \text{ (m)}^2 = 2.25 \times 10^{22} \times 12.5 = 2.8 \times 10^{23} \text{ m}$$
$$\sigma = 5.67 \times 10^{-8} \text{ W}$$
$$T^4 = 3000^4 = 8.1 \times 10^{13} \text{ K}$$

What about 2.8×10^{23} m $\times 5.67 \times 10^{-8}$ W $\times 8.1 \times 10^{13}$ K $= 1.2 \times 10^{30}$ W?

This is a luminous star! It is 10,000 times more luminous than the Sun. Probably a red giant as deduced from its low temperature.

We can see what an important tool this is as it gives us a determination of what kinds of stars we are dealing with in our observations. We can determine real, physical parameters and understand what is happening to the stars and why these traits are what they are. Across the H-R diagram, the above tools give us confidence that we are understanding the mechanism behind what we observe.

Once hydrogen exhaustion takes over, stars evolve toward the red end of the H-R diagram, and as they cool and expand, their surface areas increase and they become more luminous and move toward gianthood. From the foregoing we can now move onto examining the next stages in stellar evolution – and comprehend the processes responsible for variability.

Variability After the Main Sequence

There are a number of variable patterns in stars that have left the main sequence. Stars generally become red giants and undergo long period variability or irregular variability, whereas others are attempting to return to

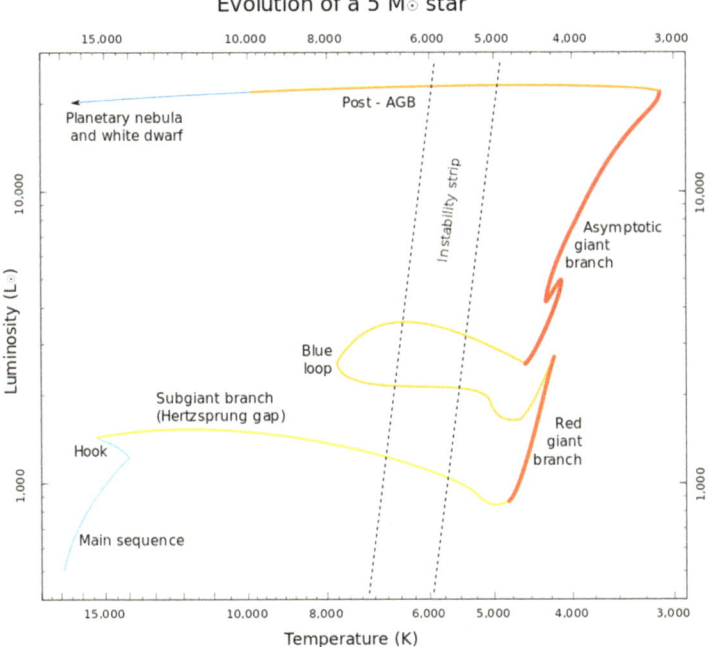

Fig. 2.8 HR Diagram of red giants

the main sequence and cross an area known as the Cepheid instability strip. In many cases, variability of these types of stars are dependent on their initial mass and therefore on where they fit on the H-R diagram. Some stars undergo various types of variability as they cross and re-cross the diagram as new forms of nuclear fusion dominate in their interiors or as mass is lost to the interstellar medium. The H-R diagram here describes the evolution off the main sequence and into red gianthood.

As stars evolve away from the main sequence, some dramatic changes occur internally that lead to variable activity. The most common mechanism for the variability of stars at this stage is known as the κ or kappa mechanism and was discovered by Harlow Shapley and independently by Arthur Eddington. The process is named after the Greek letter kappa (κ), which is the SI symbol for the coefficient of absorption (or opacity) of stellar material. The process is also known as the Eddington valve. In this case, the process drives the star to pulsate, almost literally to breathe in and out so its surface area increases and decreases. This mechanism is also known as radial pulsation. The κ mechanism is responsible for the variability of many types of stars, including Cepheid variables and RR Lyrae stars.

In the ionization zone within the envelope of a star, any small rise in density produces increased opacity, and this leads to increased absorption of energy from the stellar interior. As no radiation can pass this layer, the radiation causes the layer to heat and expand, and the layer then overshoots its normal position, and the surface area of the star increases, leading to an increase in luminosity. Eventually, the internal layer's expansion leads to a drop in pressure, density and temperature (and therefore opacity) and the stellar surface returns to its normal size, setting up the conditions for the internal layer to re-absorb radiation and block its outward flow so the expansion and cooling cycle can begin again.

Another type of variable that exhibits κ mechanism traits are the β Cephei variables. These are early B spectral type main sequence and subgiant stars that exhibit small rapid variations in their brightness due to pulsations of the stars' surfaces. These very hot stars are about 7 to 20 solar masses and are thought to pulsate due to the unusual properties of iron at temperatures of 200,000 Kelvin in their interiors. These type variables pulsate due to the kappa mechanism in a similar manner to δ Cephei variables, but the two types are distinctly different in spectral type and mass. The iron is not a part of the core but resides in the internal envelope of the star and creates a "barrier" wherein the radiation cannot escape from the star due to the opacity of the iron. The star then expands, the iron then allows the radiation through, cooling that layer, and the layer sinks toward the center of the star to start the process all over again. The bright star Spica in the constellation of Virgo was a beta Cephei-type variable, but its pulsations stopped in 1970, and they have not been recorded since. Other visible examples of Beta Cephei variables are β Canis Majoris, β Centauri and η Orionis.

Subgiants more massive than the Sun can also cross the Cepheid instability strip at a low point in the H-R diagram – and can do so again at later stages of their life cycle. As an example, at the other end of the scale from the β Cepheids are the stars known as δ Scuti variables. The H-R diagram here gives a good account of the types of variable stars we shall encounter in this book.

These are late A and early F main sequence and subgiant stars, and like the classical Cepheids, they can be used as distance indicators as they follow the period luminosity law. However, as they are generally relatively faint stars, then distance indication is limited to nearby galaxies and globular clusters. Delta Scuti stars exhibit both radial and non-radial luminosity pulsations. Such pulsations occur when some parts of the surface move inwards and some outward at the same time.

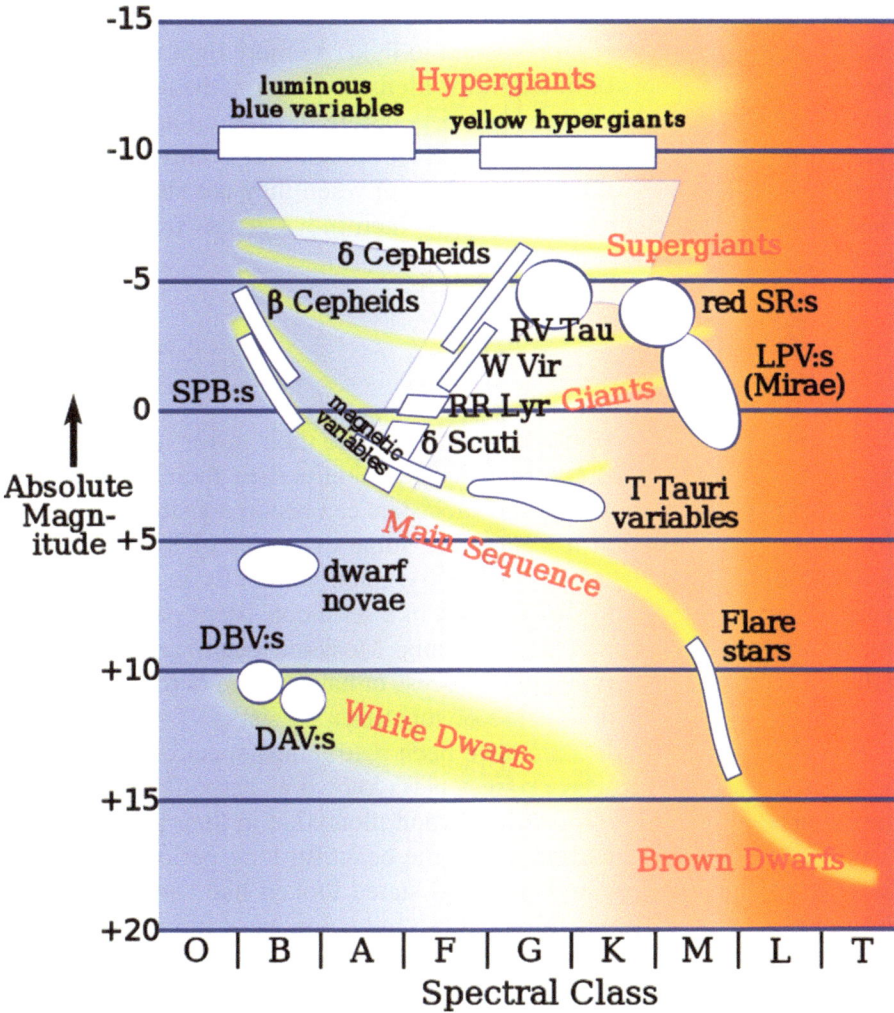

Fig. 2.9 Variable stars on the HR diagram

Again, variations in light output are due to the swelling and shrinking of the star through the Eddington valve or κ mechanism, but in this case many of these stars have been observed to have a helium-rich photosphere. As the helium is heated it becomes more ionized, which renders it more opaque to radiation leaving the star. So at the dimmest part in the cycle, where the star is at its smallest radius, the star has highly ionized opaque helium in its photosphere. The energy from this "blocked light" causes the helium to heat up and expand before the process reverses and the helium regains its electrons and thus becomes transparent to light. As more light streams from the star, the star appears brighter and, with the expansion, the helium begins to cool down. The helium then contracts before heating up again, and the cyclical process of expansion and contraction continues. Throughout their lifetime δ Scuti stars exhibit such pulsation when they are situated on the classical Cepheid instability strip. They then move across from the main sequence into the giant branch.

Variability in Red Giants

Red giant stars are extremely luminous despite their lower temperatures. As we have seen from the luminosity equation earlier, their surface area increase boosts the luminosity, and such stars can become visible over great distances. Betelguese and μ Cephei are some examples of this class, and both exhibit bouts of irregular variability. Such variability in red giants is natural; they swell up and shrink down in a semi-consistent pattern, resulting in brighter and dimmer light outputs. Most red giants are converting helium to carbon in the triple alpha process or have moved onto more exotic forms of nuclear fusion in the case of supermassive stars.

Many of these variable stars have been determined spectroscopically to be types of M spectral class or are a class known as carbon stars, as they have dominant lines of carbon and carbon monoxide in their atmospheres. A typical example of a carbon star is the beautiful long period variable R Leporis, a star the observer J. R. Hind stated looked like "an illuminated drop of blood" in reference to its intense red color. Most red variable stars observationally are more orange in color, as their chief radiative output shifts to the infrared, and the intensity of their color diminishes, as the human eye cannot see the depth of the output.

The underlying mechanism for most variability in red giants is radial pulsation, but the whole story is of course more complex than that. There are regular periods, semi regular periods and irregular periods of variability that arise due to the internal harmonics of the envelopes of red giants. As such stars are convective throughout, areas of the envelope are rising and falling at intervals that are out of step with one another, and the star literally wobbles like a jelly! Two additional types of variation have been discovered in RGB stars. One group, known as long secondary periods, show larger amplitudes with periods of hundreds or thousands of days overlying the usual variability of the star. The second variation is known as ellipsoidal variation, which is more regular.

The causes of the long secondary periods and ellipsoidal variation is currently unknown, but it has been proposed that they are possibly due to interactions with low mass companions in close orbits. In the case of ellipsoidal variations, the stars are in contact binary systems where distorted stars cause strictly periodic variations as they orbit within the tenuous envelopes of the swollen red giant. Red giants of this type therefore exhibit the characteristics of both intrinsic and extrinsic variables.

In addition, there are red giants that are going through their lives as red giants for a second time! Having been on the red giant branch, they crossed back toward the main sequence on the horizontal branch before returning to red gianthood as their fuel became exhausted. These stars are known as asymptotic giant branch stars (AGB) and appear observationally as bright red giant stars with luminosities thousands of times greater than that of the Sun.

Their interior structure of such a star is almost like an onion and is characterized by a largely inert core of carbon and oxygen, overlaid with a shell where helium fusion forms carbon that is in turn overlaid with a shell where hydrogen is undergoing fusion and forming helium. From an external perspective, the outer envelope has a material composition similar to main sequence stars. Such stars are also losing mass to the interstellar medium and are usually surrounded by large, diffuse circumstellar envelopes and are on their way to becoming planetary nebulae. Such stars often display long period or irregular variability, too. Mira is a classic example of such an AGB star with a long period of variability over 332 days.

Beyond the Red Giant Branch

There is also variability after the red giant branch. Once helium fusion begins, the star has two internal bands of energy; fusion of hydrogen happens in a shell around the core, while the next process, helium fusion, occurs in the core. The helium converts to carbon in the triple alpha process, and the carbon then further combines with helium to make oxygen. This results in the core of the star becoming rich in carbon and oxygen nuclei, and the star's surface temperature rises. On the H-R diagram the star now moves to the left to become a horizontal branch star. Red giants move to the horizontal branch of the H-R diagram and take up station as A- or F-type stars.

However, unlike main sequence A and F types, some of these stars vary in luminosity by up to half a magnitude and exhibit characteristics almost like a mini-Cepheid variable. Such stars are known as the RR Lyrae variables. These variables come under a broad heading of "cluster types," as they are generally found in globular clusters (over 80% of their class are within globular clusters) and were first discovered by Williamina Fleming in 1899. Like their larger brothers the classical Cepheids, they also follow the period luminosity law discovered by Henrietta Leavitt. However, unlike regular Cepheid variables, RR Lyrae types do not follow a strict period-luminosity relationship at visual wavelengths, although they do in infrared. RR Lyrae stars are generally about the same mass as the Sun. For a short time, low mass stars will be RR Lyrae stars on the instability strip, whereas the high mass stars have another destiny.

Lying above the RR Lyrae types on the H-R diagram are stars known as the classical Cepheids. These are stars that undergo pulsations regularly, usually with periods ranging from a few days to a few months depending on their mas. Typically, classical Cepheids are between 4 and 20 times more massive than the Sun and can be up to up to 100,000 times more luminous. They are yellow giants and supergiants, generally of spectral classes F6 to K2, and are on the horizontal branch of the H-R diagram crossing an arc known as the Cepheid instability strip, as can be seen from Fig. 2.10.

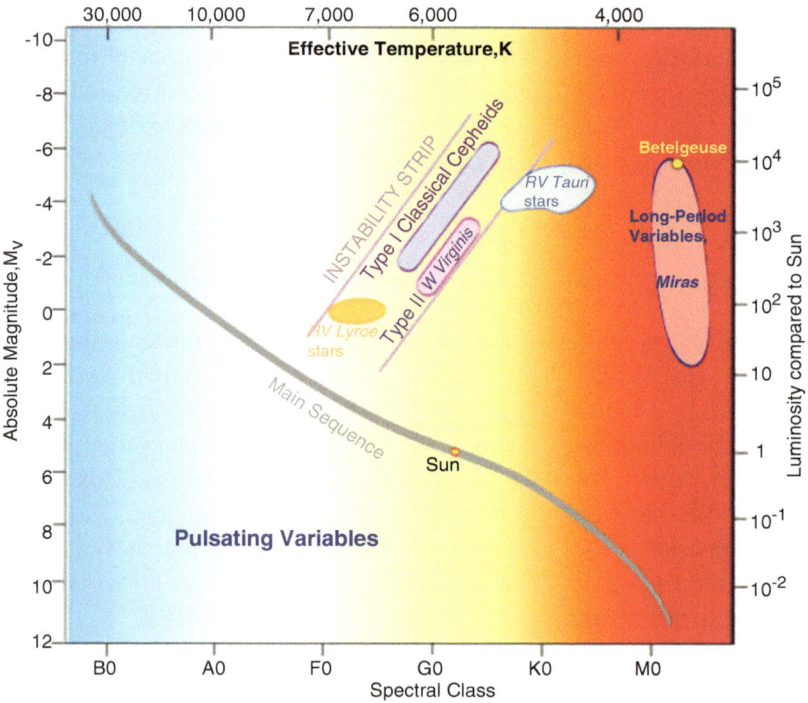

Fig. 2.10 Cepheid stars (Image from https://sites.google.com/site/aavsosppsection/spp-star-classes.)

Classical Cepheids are huge stars and undergo tremendous changes. Their radii can change in the course of their pulsations by over 25%. In a giant star like this, such a figure represents a change of millions of kilometers during a pulsation cycle!

After existing as horizontal branch stars for a few million years, the helium in the core of the Cepheid becomes exhausted, and a core full of carbon and oxygen nuclei predominate, with a helium burning shell developing underneath the hydrogen burning shell. At this stage, the star expands and cools again and becomes an asymptotic giant branch star. However, as classical Cepheid stars are so important as distance indicators while on the instability strip, it may be instructive to see how Cepheids are used to do this.

The Harvard astronomer Henrietta Leavitt discovered in 1908 that there was a relationship between how bright these stars were and the period over which they varied in luminosity. This startling relationship opened a new view on the heavens. The relationship can be expressed as:

$$M = -2.78 \times \log(P) - 1.35$$

where P = period of variability in days and M = absolute magnitude.

As the apparent magnitude will be gauged from observation, determining the distance becomes a simple matter of distance modulus then:

$$D = 10^{(m-M+5)/5}$$

Consider a star with a period of 7.9153 days. The period luminosity relationship is:

$$-2.78 \, (\log P) - 1.35$$

$$\mathrm{Log} \, P \, (\log \text{ of } 7.9153) = 0.898:$$

$$-2.78 \times 0.898 - 1.35 = -3.84$$

The absolute magnitude is −3.84.

From there it is then a simple distance modulus to get the correct distance. If the apparent magnitude is 17.7 (based upon visual observations of the mean magnitude throughout the period) and the absolute magnitude based on the above is −3.84:

$$D = 10^{(m-M+5)/5}$$

$$17.7 - 3.84 = 21.54$$

$$21.54 + 5 = 26.54$$

$$26.54 \div 5 = 5.308$$

$$10^{5.308} = 203,235 \, \text{parsecs}$$

A parsec is 3.26 light years, so:

$$203,235 \times 3.26 = 662,548 \text{ light years}$$

This relationship was first used to work out the distances to globular clusters and then to galaxies!

The brightness of any Cepheid can also be measured by using $2.512^{(m-m)}$. So if a Cepheid has a peak magnitude of 1.7 and a minimum magnitude of 4.3:

$$4.3 - 1.7 = 2.6$$
$$2.512^{2.6} = 10.9$$

The Cepheid is 10.9 times brighter at maximum than minimum!

However, the above stars are continuing to evolve away from the main sequence. What happens to large stars when they end their lives?

Supernovae

The death of a massive star is a relatively rare event. Nevertheless, there are enough of these rare but exciting objects to become worthy of study, and they generally give themselves away due to the expulsion of materials in shells or nebulous clouds (Wolf-Rayet stars), have large UV excess or are luminous blue variables (LBV). Stars of O and B spectral types with the MKK classification of Ia or Ib added to their spectral class are also noteworthy. We now know that when a luminous blue star of at least 20 times the mass of the Sun goes through its lifetime, it passes through the hydrogen fusion stage in just a few million years before going on to the helium, carbon and silicon burning stages. The tracks on the H-R diagram of such a star from birth to death can be seen in Fig. 2.11.

Once the silicon is turned to iron in the core the last (exothermic) process that holds the star up against gravity is over. To make iron fuse into the next generation of heavier elements it is necessary to inject energy into the star, as the process is endothermic – it needs energy just to keep going.

No further energy is available at this stage, and so the core falters, becoming squeezed by the overlying layers. Even this does not give the core enough energy to generate the fusion of iron, and at this point the pressurized core breaks down into simpler elements with a massive flood of neutrinos that rob the core of stability. It is no longer able to withhold the pressure from the overlying layers. The core implodes as it is robbed of energy, and the infalling layers in turn are met by a shockwave from the imploding core. At this point, a process known as explosive nucleosynthesis

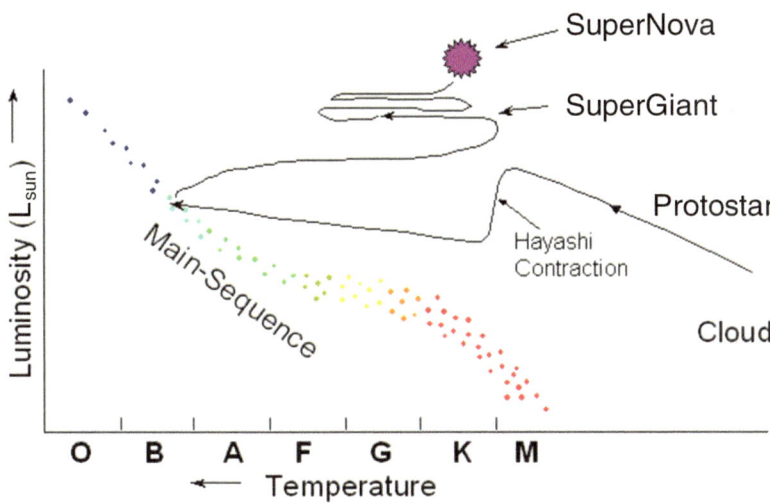

Fig. 2.11 Life cycle of a massive star (Image from https://www.pinterest.co.uk/pin/524387950353688894/.)

causes the formation of elements heavier than iron on the periodic table. The shockwave travels through the outer layers of the star, ripping them apart until the wave reaches the outer surface, where it blows the star apart. At this point, the star can become so bright that it can even outshine an entire galaxy. The materials expand away from the supernova at amazing speeds—they have been measured at 50,000 km/sec or more.

The expanding gasses left over by the explosion may be lit by internal radiation for a few months. Their light curves show a sharp decline over a few days, followed by the gradual decline in light output due to the conversion of Ni^{56} to Fe^{56}. Eventually, all that is left is an expanding patch of gasses. The core of the star, if anything is left over, may become a black hole or more likely a neutron star, such as the one in the Vela supernova remnant and in the famous Crab Nebula, Messier 1.

Parameters of Eclipsing Variable Stars

Eclipsing variable stars are merely ordinary stars that are in binary systems. They are therefore subject to the same occurrences and evolution as we have seen above. However, if we have two disparate stars in a binary system, their evolutionary histories will be different.

The wonderful thing about eclipsing binary systems is that the interaction between the stars can be reduced to a light curve, and from that light curve a great deal of information can be extracted using Newtonian mechanics. We can determine the orbital parameters, the individual masses and from there, the life cycles of each of the stars in the system and infer the physical changes that will occur in future to that system.

Astronomers can perform a full analysis via spectroscopy that will measure the radial velocity to reveal the masses of the stars and the overall scale of the system. To enable this, a typical requirement for an eclipsing system is thirty spectra of sufficient quality to reliably detect the spectral lines from both stars in the system. Detached eclipsing binaries such as Algol are easy to obtain information of this type from, and much is gleaned from studies of the contact binary and semi-detached stellar systems exhibited by some eclipsing variables.

As the stars age, they begin to expand, and the gravitational effects upon the system begin to change. Once a star fills a gravitational parameter known as the Roche lobe, materials from that star will begin to be transferred to the companion star. The "dragging" on the secondary star then moves it closer to its larger companion, and mass transfer begins to increase and thereby alters the spectral characteristics of both stars. Eventually, enough materials will be lost to enable the companion star to become hotter and shift from the spectral range of G or F to the A spectral type, while the original larger star becomes K and M type.

The change in orbital dynamics of the system and its subsequent evolution depends on the stellar masses, the exchange of materials and the timescale of the exchange. In many cases astronomers observe the less massive star in some systems, an evolved subgiant, transferring mass in a stream down to the more massive one on the main sequence. The stars themselves have now changed with the original larger star now an evolved and less massive giant, with a bright main sequence companion that has now added the lost mass of the primary star and is going through its evolution faster due to mass transfer.

These supergiant main sequence pairs potentially become cataclysmic variable stars. As the supergiant's expanding outer atmosphere is drawn toward the smaller star, the gaseous exchange will start to slow the orbital system down, allowing the rotational period to shrink and the orbits of the stars to get closer and closer. The envelope of gas surrounding the stars is ejected by stellar winds from the larger star until in many instances all that is left of the supergiant is a stellar core, which becomes a white dwarf. The white dwarf then feeds off gasses from the secondary star until there is buildup of materials on the surface, which become compressed enough for a thermonuclear explosion to occur. What we see as external observers when this occurs is the sudden brightening of the system in a nova. Beforehand, the materials accreted onto the surface of the former supergiant star will produce a wobbly light curve indicative of a rotating hotspot and an accretion disc. Such systems give out soft X-rays, and many such pre-nova cataclysmic variables have been catalogued.

Eclipsing or close binary partners therefore lead to changes in evolution and structure between the components, and their behavior can be exhibited, dependent on original masses, as eclipsing companions, X-ray binaries or cataclysmic variables. We shall examine some examples in this book.

Conclusion

We can certainly see from this the importance of variable stars across a wide spectrum of astronomical activity. Understanding variable stars has taught astronomers how interconnected the universe is and can place us in a position to predict what will happen to stars during their lifetimes.

All of the foregoing may sound a little academic and unnecessary in a book of this kind. However, a deeper understanding of the processes that occur out there in our galaxy contributes not only to one's understanding of the subject but also enhances the awe and wonder we experience when we look through the eyepiece or at a photograph of an object. Early astronomers had no idea of what sort of objects they were viewing, and subsequent visual observers of variable stars could only describe the basics of the light curves; they had no idea of the true nature of the objects under review. Today we have a very good understanding of the internal constitution of the stars that really is a triumph of 20th century astrophysical ingenuity. Don't dismiss this greater understanding as irrelevant to our discussions, as knowing the details only adds to your experience of the night sky.

Now that we have discussed the history and astrophysical understanding of variable stars, we need to make preparation to observe these objects in their correct setting under the night sky. What techniques are required to observe them in safety and comfort to gain the maximum from our observing sessions? Let us now turn to a discussion of this topic.

Chapter 3

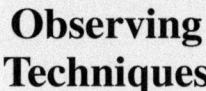

Observing Techniques

Techniques for observing variable stars are very similar to any other types of observing. We require the same tools, patience and dedication as any other field. However, it must be acknowledged that observing variable stars requires a few techniques that are quite different to normal observing methods and necessitate some measure of astronomical experience and background. Besides being able to estimate the magnitudes of stars to a high degree of accuracy, modern variable star observing also demands a knowledge of photometry, computing, reduction to light curves via online software, or making one's own light curve by means of *MS Excel* or similar and a host of extra knowledge and methodologies that are part and parcel of modern astronomy.

However, there are some basic techniques that everyone would do well to keep in mind, and we shall examine these here. These techniques are taken from my book *Observing Nebulae* and the following chapter follows almost verbatim the advice given there, should anyone wish to cross reference these techniques.

Seeing and Transparency

To see deep sky objects and faint stars at their best, avoid times of the month when the Moon is shining brightly. Although the Moon is a lovely romantic object shimmering with a silvery light, looking wonderful in a cloudless

© Springer Nature Switzerland AG 2018   47
M. Griffiths, *Observer's Guide to Variable Stars*, The Patrick Moore
Practical Astronomy Series, https://doi.org/10.1007/978-3-030-00904-5_3

sky, it is less than romantic to astronomers and photographers interested in digging out remote or obscure nebulae. During an average month there is at least a two-week Moon-free period, when deep sky objects can be seen at their best. In addition, only observe when the air is fairly clear and the atmosphere, or seeing conditions, remain quite steady. High, hazy clouds do rather spoil the view and renders faint nebulae invisible.

Thus, seeing the sky at its best takes a little patience. Even the most perfect summer day can affect the observing conditions at night as the heated atmosphere gives rise to tremulous effects that then have an impact on the astronomical "seeing," a term that describes the properties of the atmosphere. Turbulence from heat rising and humidity all detract from the perfection of crisp point-like stars, so estimating the seeing is actually quite important – especially if one is going to examine binary stars. Some close doubles will not be split under conditions of bad seeing. Added to this problem is that of transparency, a term used to describe the clarity of the sky. How does one estimate the effects of seeing and, equally important, what is the transparency factor?

The astronomer Eugene Antoniadi invented a "seeing" scale that has become the most dominant model in use. The important thing to note is that seeing is usually defined as the steadiness of the air. The Antoniadi scale is:

I.   Perfect seeing without a quiver
II. Slight quivering of the image with perfect moments of calm for several seconds
III. Moderate seeing with larger air tremors that blur the image
IV. Poor seeing. Constant troublesome undulations that spoil the image
V.   Very bad seeing, hardly enabling a sketch to be made

This scale helps the observer to make an observation of how calm the sky is to enable him or her to take an image or to physically observe – or not observe at all!

Transparency is affected by two things: darkness and extinction. Darkness is obviously a measure of how dark the night sky is and is affected by such things as twilight, moonlight and light pollution and is tied to the scattering of light particles in the atmosphere by the air itself or by droplets of water vapor, smoke or dust. Where the atmosphere is unaffected by haze or humidity, the transparency will be excellent. A transparency scale with letters can be utilized here so a typical example will be:

A   Dark blue sky – excellent transparency
B   Medium blue sky – Above average transparency
C   Light blue - Average sky transparency
D   Pale blue – Poor sky transparency
E   Gray – very poor sky transparency

Both scales when used together can be excellent indicators of the prospects for good observing, where even some of the fainter objects can stand out with good clarity under excellent skies. Some of the best observing conditions with perfect transparency and seeing is enjoyed during the cold winter months, when the air is steady and clear after a frosty day with little humidity and no rising warm air.

However, all of this would be rendered moot unless one has access to a dark sky site or some way of classifying your location so that limiting magnitudes, perceptions of transparency and a host of other considerations are taken into account. A handy tool that pulls these concerns together is the Bortle scale.

The Bortle Scale

The Bortle scale is a numeric scale that measures the brightness/darkness of the night sky from any location. In some respects it gives a quantifiable record of the visibility of celestial objects and the interference to such caused by light pollution. The scale was created by the observer John Bortle and published in the February 2001 edition of *Sky & Telescope* in an attempt to aid amateurs in evaluating the darkness of an observing site and provide a tool to compare the darkness of observing sites. The scale ranges from Class 1, the darkest skies available on Earth, through Class 9, inner-city skies. It gives several criteria for each level using naked-eye limiting magnitude (NELM) as a guide. Sky conditions, depending on magnitude, can then be judged by the Bortle scale, now a recognized dark sky determination tool, though recently the accuracy of the scale has been questioned. Nevertheless, it provides a fair ready reckoner for sky quality.

Visual estimates of stars and other deep sky objects for levels of transparency and sky clarity can also made. The globular clusters Messier 13 (Mag 5.8), Messier 3 (Mag 6.2) and Messier 5 (Mag 6.6) should be clearly visible with the unaided eye, and even the galaxy Messier 81 (Mag 6.9) should be visible with a little averted vision. If you use a sky quality meter as noted above, then the readouts can be converted into Bortle scales by means of the following chart 3.2:

If your skies are Bortle Class 3 – a suburban rural transition – that is a good sign of exceptional skies. This is one reason why as a quick guide the Bortle scale is used in siting small astronomical observatories.

Now that magnitude has been mentioned, what exactly is it? Magnitudes indicate the brightness of individual objects in space no matter if they are stars, nebulae, galaxies or planets. It is an astronomical estimate of how

Class	Color Key	Naked–eye Limiting Magnitude	Sky Description	Milky Way	Astronomical Objects	Zodiacal Light / Constellations	Airglow and Clouds	Night Time Scene
1		7.6 – 8.0	Excellent, truly dark–skies.	MW shows great detail and light from the Scorpio / Sagittarius region casts obvious shadows on the ground.	M33 (the Pinwheel Galaxy) is a obvious object.	Zodiacal light has an obvious color and can stretch across the entire sky.	Bluish airglow is visible near the horizon and clouds appear as dark blobs againt the backdrop of the stars.	The brightness of Jupiter and Venus is annoying to night vision. Ground objects are barely lit and trees and hills are dark.
2		7.1 – 7.5	Typical, truly dark skies.	Summer MW shouws great detail and has veined appearance.	M33 is visible with direct vision, as are many globular clusters.	Zodiacal light bright enough to cast weak shadows after dusk and has an apparent color.	Airglow may be weakly apparent and clouds still appear as dark blobs.	Ground is mostly dark, but objects projecting into the sky are discernible.
3		6.6 – 7.0	Rural sky.	MW still appears complex, dark voids and bright patches and meandering outline are all visible.	Brightest Globular Clusters are distinct, but M33 is only visible with averted vision. M31 (the Andromeda Galaxy) is obviously visible.	Zodiacal light is striking in Spring and Autumn, extending 60 degrees above the horizon.	Airglow is not visible and clouds are faintly illuminated except at the zenith.	Some light pollution eveldnet along the horizon. Ground objects are vaguely apparent.
4		6.1 – 6.5	Rural / suburban transition.	Only well above the horizon does the MW reveal any structure. Fine details are lost.	M33 is a difficult object, even with averted vision. M31 is still readily visible.	Zodiacal light is clearly evident, but extends less than 45 degrees after dusk.	Clouds are faintly illuminated at the zenith.	Light pollution domes are obvious in several directions. Sky is noticeably brighter than the terrain.
5		5.6 – 6.0	Suburban sky.	MW appears washed out overhead and is lost completely near the horizon.	The oval of M31 is detectable, as is the glow in the Orion Nebula.	Only nints of zodiacal light in Spring and Autumn.	Clouds are noticibly brighter than the sky, even at the zenith.	Light pollution domes are obvious to casual observers. Ground objects are partly lit.
6		5.1 – 5.5	Bright, suburban sky.	MW only apparent overhead and appears broken as fianter parts are lost to sky glow.	M31 is detectable only as a faint smudge; Orion Nebula is seldom glimpsed.	Zodiacal light is not visible. Constellations are seen and not lost against a starry sky.	Clouds anywhere in the sky appear fairly bright as they reflect back light.	Sky from horizon to 35 degrees glows with grayish color. Ground is well lit.
7		4.6 – 5.0	Suburban / urban transition.	MW is totally invisible or nearly so.	M31 and the Beehive Cluster are rarely glimpsed.	The brighter constellations are easily recognizable.	Clouds are brilliantly lit.	Entire sky background appears washed out, with a grayish or yellowish color.
8		4.1 – 4.5	City sky.	Not visible at all.	The Pleiades Cluster is visible, but very few other objects can be detected.	Dimmer constellations lack key stars.	Clouds are brilliantly lit.	Entire sky background has an orangish glow and it is bright enough to read at night.
9		4.0 at best	Inner city sky.	Not visible at all.	Only the Pleiades Cluster is visible to all but the most experienced observers.	Only the brightest constellations are discernable and they are missing stars.	Clouds are brilliantly lit.	Entire sky background has a bright glow, even at the zenith.

Fig. 3.1 The Bortle scale

Color Magnitude	Bortle Class	Sky Brightness mag/arcsec²	Artifi./Natural
7.6 - 8.0	1	>21.90	<0.01
7.1 - 7.5	2	21.90 - 21.50	0.01 - 0.11
6.6 - 7.0	3	21.50 - 21.30	0.11 - 0.33
6.3 - 6.5	4	21.30 - 20.80	0.33 - 1.00
6.1 - 6.3	4.5	20.80 - 20.10	1.00 - 3.00
5.6 - 6.0	5	21.1 - 19.10	3.00 - 9.00
5.0 - 5.5	6,7	19.1 - 18.00	9.00 - 27.0
<4.5	8,9	<18.00	>27.0

Fig. 3.2 Bortle SQM chart

bright each object can appear in the sky or would appear if artificially brought to a useful distance for comparison. Some nebulae are large and have integrated magnitudes that make them appear bright on paper, but in reality they are often amorphous and a little lost against the background sky. Therefore, some knowledge of the magnitude scales and their use is helpful for the observer.

Magnitudes and True Brightness

The magnitude scale is the astronomical standard for measuring the brightness of any object. Stellar magnitudes are measured using a logarithmic scale in which each magnitude differs from the next (brighter or fainter) by 2.512, or 2.5 times as bright/faint. The standard star in astronomical use for calibration is Vega (alpha Lyrae) with a magnitude of 0.0. In astronomical measurement of magnitude, a brighter object has a lower number while fainter objects have higher numbers. To determine the differences between bright and faint stars the following table can be used:

Magnitude	Brightness change
1	2.5
2	6.25
3	16
4	40
5	100

So, a change in 5 magnitudes = change in brightness of 100. As an example of how this works in practice, Star A has an apparent magnitude of 2. Star B has an apparent magnitude of 4. How much brighter is Star A than Star B?

$$\text{Magnitude difference}: 4 - 2 = 2$$

So, Star A is 6.25 times brighter than Star B

Again, to see how this works in practice we can determine that a star may have an apparent magnitude of 1.7 and another has an apparent magnitude of 7.6. So:

$$7.6 - 1.7 = 5.9$$
$$2.512^{5.9} = 229 \text{ times brighter}$$

Similarly, stars of magnitude 14.6 and 3.8:

$$14.6 - 3.8 = 10.2 \quad \left[2.512^{10.2}\right] = 12028$$

Stars with widely different apparent magnitudes:

$$17.5 \text{ and } 1.7 \quad 17.5 - 1.7 = 15.8 \quad \left[2.512^{15.8}\right] = 2,090,789$$

And a final example shows how faint some objects can become! Betelguese has a magnitude of 0.4. The faintest objects in the Hubble deep field are mag 31.

$$31 - 0.4 = 30.6 \quad \left[2.512^{30.6}\right] = 1.7 \times 10^{12}$$

Really faint then! The magnitude scale is essential in understanding the brightness of objects relative to each other and gives the observer some hint of the kinds of objects visible in their telescopes. This is very important when one wants to see variable stars visually, as it is important to make visual comparisons with other, stable stars in the same field of view.

Dark Adaptation

Understanding the importance of transparency, seeing and the Bortle scale would be rendered moot if we ignored the simple tool that all observers rely on, regardless of their activities. Under a dark sky, take a little time to allow the pupil of your eye to open fully. This technique is known as dark adaption. This process can take about five to 10 minutes, but to be fully dark-adapted takes up to 30 minutes. During this period of time, avoid lights of any kind; do not switch on your torch or stare at nearby street lighting.

Once this has been accomplished, you can begin observing, as your eyes are now a little bit more sensitive to light than they ordinarily are during the day. However, some astronomers have noted that faint objects seem to be a little brighter if they are seen "out of the corner of the eye," as it were, and this phenomenon is an important tool of the observer. The process of seeing objects in greater detail simply by not looking directly at them is called averted vision.

This phenomenon arises due to the way that the eye is constructed. At the rear of the eye is a light-sensitive membrane, the retina. The retina is fabricated from two sets of cells, commonly called rods and cones. The cones lie directly behind the pupil and receive most of the incoming light, but the light-sensitive rods lie off to the sides of this aperture. Thus, when not looking directly at some object, the more sensitive rods are able to pick up the stray light that the cones are missing, making averted vision a good habit for astronomy.

To maintain your night vision, it is best to examine any star charts or atlas by means of a torch with a red beam. This red light will not interfere with a dark-adapted eye, and is comfortable and easy to see such charts by. When observing, make sure that your comfort is the paramount consideration, so always wrap up warm, have a hot drink handy and take a break every hour or so if you intend to observe all night. Additionally, if you can stand on a raised board while observing, then the heat of your body will not be sapped through your feet, leaving you cold and miserable. This is simple common sense, yet many observing sessions have been ruined by the lack of such preparation.

Additional Materials for Variable Star Observing

Recording the magnitude output from a variable star is a skill that develops with time and practice. However, there are some techniques that are essential to your success in recording accurate measurements of the light output, and not all of these are down to instrumentation.

Red Flashlight

One of the most obvious yet overlooked tools is a red light flashlight. A red light torch is very handy for setting up your telescope, reading star charts and finder charts and of course entering any notes of the variable star in your observing log. A red head torch or an ordinary white light torch with red paper over the lens will do very well.

A Watch or Clock

It is important to record the time as accurately as possible when observing variable stars. Some stars, such as cataclysmic variables, vary over such short periods that accurately timing their rise and fall in brightness is essential. Many observers have Fitbits™ or watches that are visible in the dark or in dim light, and even smartphones are used as simple yet accurate timepieces. No matter what the observer uses, remember that all times should be recorded in UT (GMT) and to the nearest minute.

Notebook for Recording

A notebook for recording is essential, as it is not always possible to remember every detail about your observation. The more accurate your recording, the better your observations will be. To that end, it is recommended that several items should be recorded in your notebook: the date and time of the observation; the star observed; the estimate, either fractional or Pogson step and the class of observation, which is a little subjective and dependent on each observer such as:

A) Very confident observation made in ideal conditions,
B) An observation with average confidence – maybe due to moonlight or hazy cloud
C) An observation of low confidence, but still worth recording.

Finally, note the local weather conditions, record any problems with light pollution, record the phase of the Moon (if visible) and don't forget to record the instrument used in the observation.

Variable Star Chart

It is best to use an official chart when locating and observing any variable as far as you possibly can. Charts are usually available from the AAVSO or the BAA. The comparison stars listed on the chart have very accurately measured magnitudes and have been selected to reflect a suitable star color that contrasts with the variable. The right ascension and declination of the star will also be recorded on the chart, and in some cases basic information about the star's type and magnitude range will also be included.

If you are looking at a range of variable stars, then knowing the limiting magnitudes of your equipment is essential. The following table from the British astronomy magazine *Sky at Night* may be a useful guide to know, as there is no point in trying to see stars at their minimum if they are beyond the reach of your instruments.

Limiting Magnitudes for Various Optical Instruments

Aperture	Instrument	Limiting mag.
10 × 50	Binoculars	+9.5
15 cm	Reflector	+13.0
22 cm	Reflector	+14.5
25 cm	Reflector	+15.0
35 cm	Schmidt-Cassegrain	+16.7
40 cm	Reflector	+16.8
45 cm	Reflector	+16.7

A few other important things to consider are the star's Designation. This is the six-digit number assigned to each variable that approximates its position in the sky. For example, R Leonis' designation is 094211, which shows that the star lies at 9 hours 42 minutes and 11 degrees north. To distinguish between northern and southern hemispheres, any star's declination that is underlined or is in italics represents south.

In addition, variable stars all have names or are catalogued according to their types, as recognized by the AAVSO and the *General Catalogue of Variable Stars*. Traditionally, all variable stars in a constellation start with the letter R and go through Z and are named in the order of discovery. After Z, it goes RR, RS, RT, SS, ST, and SU through to ZZ. After that, the designations run from AA, through to AZ, then BB to BZ and so on, but remember that the letter J is left out. As we have seen beforehand, when a constellation has so many variable stars that it exceeds these 334 letters, the next variable star becomes a number: V335, with the following becoming V336 and so on. Variable stars that have a long history of observation and Bayer designations will not be included in such nomenclature. Stars like Algol, Betelguese or Mira already have Greek letters to signify them.

All timed observations should be made according to the Julian day (JD). This is the number of the day from January 1, 4713 B. at 12:00 noon. So for example, as the day of writing, 18th December 2017, and a time of 10:30, the Julian date is: 2458105.937500. A useful online tool to find your exact JD can be accessed at: http://aa.usno.navy.mil/data/docs/JulianDate.php.

If you use the AAVSO online data entry system, you can enter Universal Time or the Julian Date. The AAVSO helpfully provides a Julian date calendar each year.

Firstly, it must be admitted that finding variable stars is not as simple a task as it first appears. Remember that occasionally the variable may be invisible at the time! It is therefore important to get to know the star fields as well as possible and compare them to the charts that are available from the AAVSO or the BAA. Needless to say that observing variable stars is best when the star is visible and one is familiar with the position of the variable star on a chart and its appearance in the star field.

Most charts, as can be seen from the BBA ones supplied with this book, already have comparison magnitudes; these are the result of many years of work by various observers, and these prepared charts have highly accurate magnitudes that have been confirmed by photometric measures. The observer will notice that the magnitudes on these charts are expressed in whole numbers, without a decimal point, so as not to confuse the observer with dots that may be misinterpreted as stars.

There are two systems in visual use by many variable star observers. These are the fractional method and a method known as the Pogson step. Both methods are well known to variable star observers and can be found in various books. The methods outlined here are taken from an amalgam of sources such as *Observing Variable Stars, Novae and Supernovae* by Gerald North, *Observing Variable Stars* by David Levy and the *Webb Society Deep Sky Observers Handbook* on variable stars. Other sources are available online, especially Karen Holland's excellent introduction to variable star estimation available at: http://www.britastro.org/vss/holland.pdf

The fractional method relies on estimating the brightness of the variable star between two comparison stars of known brightness. The brightness is then recorded as a fractional difference. Once the observer is familiar with these steps, the differences between the comparison stars should not exceed more than about 0.5 magnitudes, with the magnitude difference given as fractions. For example, a magnitude difference of 0.5 would result in a fraction of ½; for a magnitude difference of 0.25, the difference will be ¼ and so on.

If a variable has the unknown visual magnitude, then a comparison star in the same field could have X magnitude and another, fainter comparison star could have Y magnitude. Once the variable is identified, then estimate its brightness in comparison to X and Y. If a variable was estimated to be 1/5 of the magnitude difference between X and Y, then one can record:

$$X (1) V (4) Y$$

Note that in all aspects of fractional observing, the brighter star is always written first, and then all other stars in descending order of magnitude. As one can envision, this kind of technique demands time, patience and experience.

If your variable star is exactly the same luminosity as any comparison star, then the record should show

$$V = X$$

It is always best to check your observations against several comparison stars wherever possible, and if you can, select a comparison star that matches the color of the variable you are studying.

On rare occasions, a variable will not have a magnitude that can be compared between two stars on the finder chart, or perhaps there is only one star of a similar brightness in the field. If this is the case, then it is best to choose two stars that are not as bright or faint as the variable star, ensuring that the magnitudes of the chosen stars bracket the magnitude of the selected variable. If the variable star was is fainter, then record:

$$X \; (1) \; Y \; (4) \; V$$

This indicates that the variable is brighter than Y but fainter than X. If the estimate is from a brighter star, it can then be expressed as:

$$V \; (1) \; X \; (4) \; Y$$

This example indicates that the variable is brighter by 1/5 the amount that X is brighter than Y. If we suppose the estimate was really 1/5 the light between X and Y, then the difference also becomes 4/5 the interval between Y and X. The total of these two in terms of difference, then, remains exactly 1.00, which it must always be. We can then calculate the magnitude of our variable by making these comparisons:

$$X \; (1) \; V \; (4) \; Y$$

If X = 6.45 and Y = 7.77, then:

$Y - X = 1.32$

$1/5 = 0.26$

$4/5 = 1.06$

$6.45 + 0.26 = 6.71$ (We add here because the variable is fainter than X.)

$7.77 - 1.06 = 6.71$ (We subtract here because the variable is brighter than Y.)

The fractional method gives a fairly good approximation of the variable star's magnitude in comparison to known stars. The second method many variable star observers use is known as the Pogson step method.

This method estimates brightness in 0.1 magnitude divisions using as many comparison stars as possible. Typically, the observer selects one star being brighter and one being fainter than the variable almost as in the fractional method above. This has the advantage of detecting variability using either the comparison star or any incorrectly stated magnitude. It also means that only one comparison star can be used, especially if no other suitable star is available.

The key to this methodology is that the eye must be trained to make estimates to tenths of a magnitude. To learn this method, one has to first be able to recognize magnitude differences of 0.1, 0.2, 03, etc.; this requires patience, dedication and experience. Estimates between stars with a difference of 0.5 magnitudes are deemed too difficult for visual estimates. Many advanced and experienced observers have been so well trained that they can estimate as low as 0.05 magnitudes, though this is relatively rare!

Once you have this kind of ability, then one can estimate the magnitude of the variable star in a similar way to the fractional method.

Let us suppose that the variable is the same estimate as in the fractional method. The magnitudes are X = 6.45 and Y = 6.77, and the deduced variable's magnitude is 6.51. If other observations of comparison stars were made, in tenths of a magnitude, we can then check our accuracy by using perhaps two other stars (let us name them W and Z). These stars have the magnitudes W = 6.63 and Z = 6.3. Therefore, the magnitude estimates for the variable becomes:

$$Z - 0.2 \text{ mag.} = 6.51 \quad W + 0.2 \text{ mag.} = 6.43$$

with a mean magnitude estimate between them of 6.47.

These results state that Z was estimated by the observer to be 0.2 magnitudes brighter than the variable star while the W star was observed to be 0.2 magnitudes fainter.

One of the important things to take into account is atmospheric extinction. This is due to the differences in air mass as one observes stars either at lower latitudes of the sky or right at the zenith. This important factor can be added to your notes and calculations of magnitude to make things more accurate. As a general rule, the following table is a good approximation for atmospheric extinction, though it does not take into account the properties of the air such as cold/dryness or moisture content. Nevertheless, it should be enough for most observers no matter where they are located.

Corrections for Atmospheric Extinction

Altitude of Star Above Horizon (°)	Correction Factor (Magnitude)
60	0.04
50	0.07
40	0.13
30	0.17
25	0.23
20	0.32
15	0.7
10	1.0
5	1.7

Locating and estimating variable stars near the horizon requires larger corrections, so it's best to observe the star when it is at a greater altitude.

Obviously, getting used to doing this is going to take some time and attention, but the foregoing instructions are staple estimates for visual observation that have been in use for a long time. The advantage of such methods is that they can be assessed very quickly by the observer, and your observations checked and repeated by other observers, too. In order to enable such accuracy, it is important to take detailed notes of your observing sessions.

Drawing a Light Curve

Once you have the basic methods of estimating magnitude you can then construct a light curve for the star. This is a relatively simple graph that accounts for changes in brightness over a period of time. If you start with variable stars that vary over a matter of days, then it is easy enough to start a chart recording the data. Fig. 3.3 shows a typical light curve chart for a long period variable star.

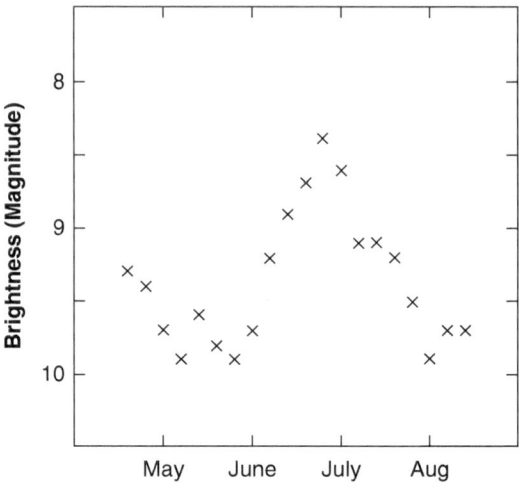

Fig. 3.3 Simple light curve (Image from https://imagine.gsfc.nasa.gov/science/tool-box/timing1.html.)

The X axis of the chart records the days (Julian time if you wish) over which the star varies, and the Y axis records the fluctuations in magnitude. To begin with one would simply place a system of points corresponding to the day and the estimated magnitude of your target. After some time of observing the star one should have a multitude of points that can be tied together with a curve going through the rise and fall in magnitude over the time period – a light curve!

Most light curves are going to be simple, but periodically the light curves can deviate from the given data – or your collected data – due to the presence of perhaps a third partner in some eclipsing star systems, or apsidal motion that throws the observer off balance. In such situations, many astronomers make what is known as O-C diagram, where the O stands for "observed" data and C stands for "calculated." If the true period of the star is longer than your calculation, then the light curve will slope upwards. If it is shorter than the true period, your calculated light curve will have a downward slope. It can be seen therefore that calculation can lead into some tricky situations where the light curve does not reflect the star's real output, but this method is still effective and can be corrected by experienced observers so that a better fit can be shown once enough data has been observed.

Observing Notes and Records

One of the most important things about observing anything in the sky, but especially for variable stars, is to record things as accurately as possible. To do that, a comprehensive observing log can be made by the observer so that all details of the session can be recorded and any activity can then be followed up in the case of any discovery, or just to ensure that your experience and light curve recording grows.

You can make an observing log yourself, but at the end of this chapter you can see one in Fig. 3.4, a typical example that this author, as the director of a small observatory, uses to ensure accuracy of results and in order to collect data on the number of observing sessions over the course of the year. Also appended as Figs. 3.5 and 3.6 can be seen the input pages of the AAVSO to make light curves from your observations or to place your visual observations online as a permanent record.

As can be seen, planning your observations is crucial, and you must know the times that your objects will be visible at best. Include any information you require in advance in this section and then augment the log with your notes. Be sure to record as much detail as possible. If you are imaging, then record the images taken to ensure that you have good notes in case of discovery, especially for supernova patrol. Anything else, such as sketches, field of view, variable star finder charts and comparisons can also be placed on the log. It's up to the reader.

We hope that the foregoing has given the potential variable star observer some food for thought and introduced the basic methods of observation. Our next chapter will discuss the various optical aids available to variable star observers before we move on to the more complex recording of variable stars by photometry.

Observing Session

Date:

Location:		Equipment:	
Event:	Group:		
Moon Phase:	Seeing/Transparency		
Limiting Magnitude	% Cloud Cover	Approximate Temp	Time UT__Other____ Begin H:__M__ End H:__M__

Notes

_____
_____
_____
_____
_____
_____
_____
_____
_____
_____
_____
_____

Plan

_____
_____
_____
_____
_____
_____
_____
_____
_____
_____
_____
_____
_____
_____

Fig. 3.4 Observing log

Observed Objects

Object Name	Type	Time	Description	Sketch/Image Reference

Fig. 3.4 (continued)

Images

Reference: _____

Notes/Description

_____
_____
_____
_____
_____

Reference: _____

Notes/Description

_____
_____
_____
_____
_____

Fig. 3.4 (continued)

Sketches

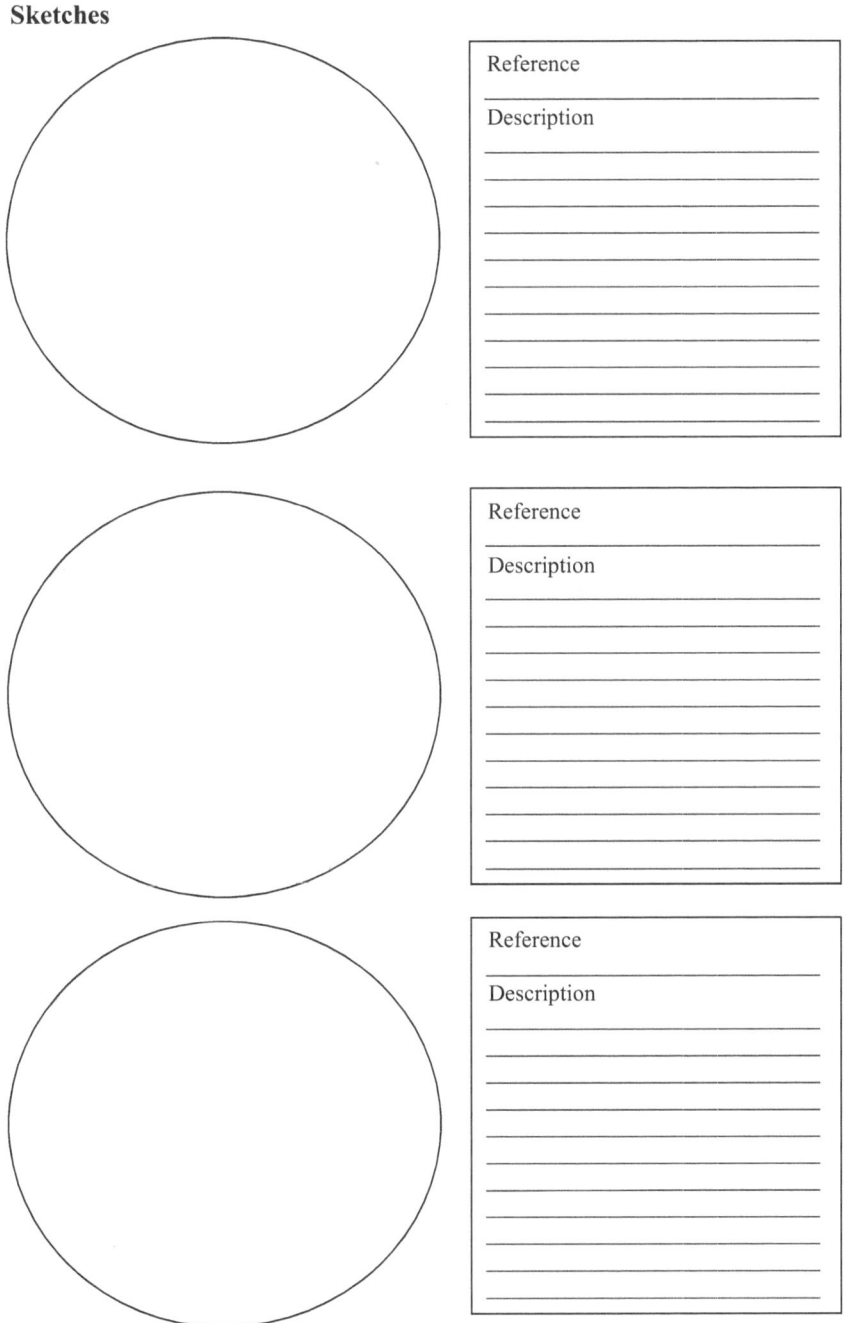

Fig. 3.4 (continued)

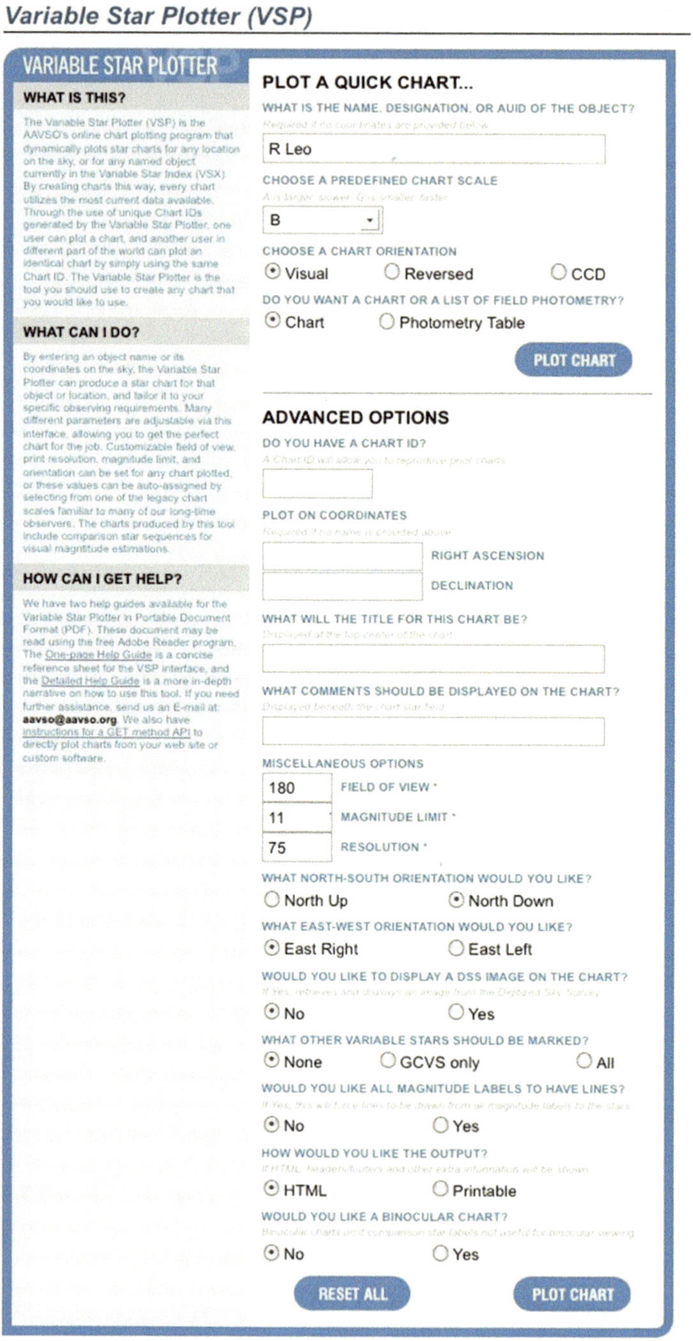

Variable Star Plotter (VSP)

Fig. 3.5 AAVSO computer entry for light curves

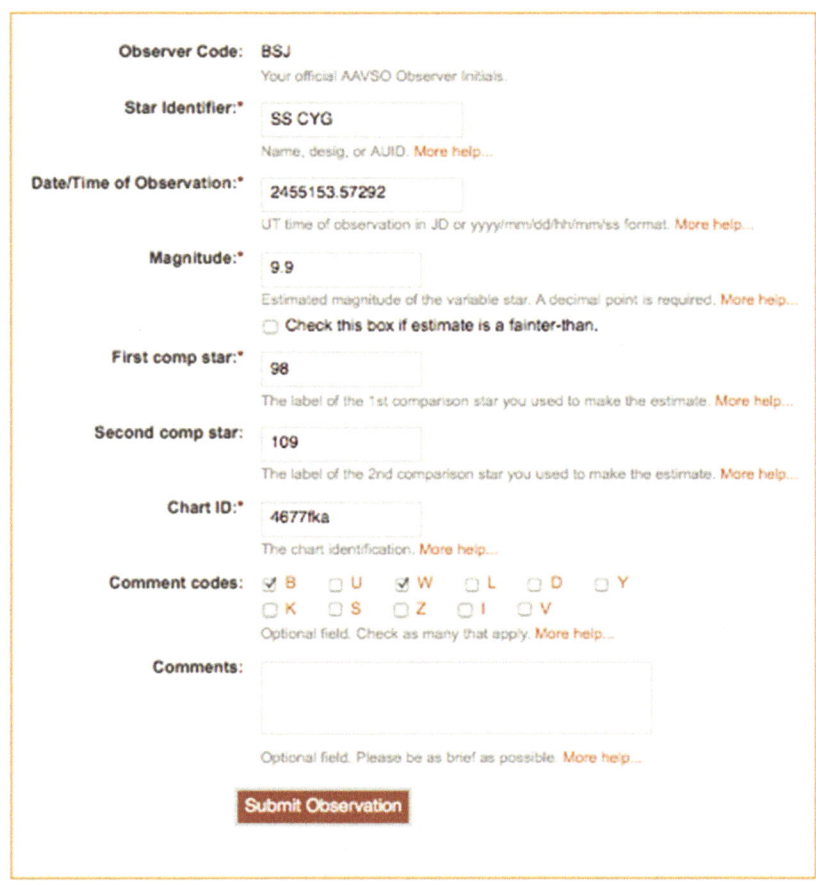

Fig. 3.6 Visual recordings entry screen

(These pages are courtesy of the AAVSO.)

Chapter 4

Instruments and Equipment

Although the night sky is available to all people, only those fortunate enough to have some additional visual aids will be able to make the most of our heavenly heritage. In our modern era it seems incredible to think that the telescope was invented only 400 years ago and the quality of astronomical telescopes has steadily improved ever since. Therefore, let us now turn to an examination of the kinds of instruments and auxiliary equipment available to sky watchers. I shall cover binoculars in the chapter about observing binocular variables.

Choosing and Using a Telescope

If you own, or wish to own, a small telescope, you must determine what's important to you. What do you most want to look at? How dark is your sky? How experienced an observer are you? How much to you want to spend? What storage space do you have, and how much weight do you want to carry? Answering these key questions and familiarizing yourself with what's on the market will enable you to acquire a telescope that will work well and satisfy your observing needs for many years. Plus you can always trade it in for a larger one!

One of the first considerations is the mount. Most small telescopes do not have very good mounts, ones that are stable and vibration free. Many of the cheaper ones are simple alt-azimuth mounts with a few screws to secure the

© Springer Nature Switzerland AG 2018   69
M. Griffiths, *Observer's Guide to Variable Stars*, The Patrick Moore
Practical Astronomy Series, https://doi.org/10.1007/978-3-030-00904-5_4

'scope, which leads to lots of trouble selecting your object and keeping it in the field of view as one tightens the screws. Even when in use, simply brushing the 'scope means that it moves off target, and acquiring your image again can be a difficult process.

If you are purchasing from a supplier, check reviews before selecting a 'scope, as a good mount is essential to your observing experience. An equatorial mount is preferred as it is a little heavier. It has to be set up properly, but once achieved only one axis is needed to move once you are locked on your target. Additionally, the advantage of equatorial mounts is that they can be driven with small motors, and they can be used as camera platforms. An alt azimuth mount is only an advantage if one is purchasing a reflecting telescope on a Dobsonian mount. For any other telescope this author considers them to be severely limited. The stargazer must make decisions based on needs and budget, but there are some excellent quality telescopes on the market that have equatorial mounts that are very good value and give a fairly vibration free image.

One of the things that anyone buying a telescope should know is how they differ in a property known as focal ratio – usually abbreviated to f. The f number of the system determines two properties, its field of view and how responsive the system is to incoming light. The focal ratio can be worked out simply, as it is merely the product of the focal length of the lens (how long is the light path from the lens to the focal point, where all the light rays

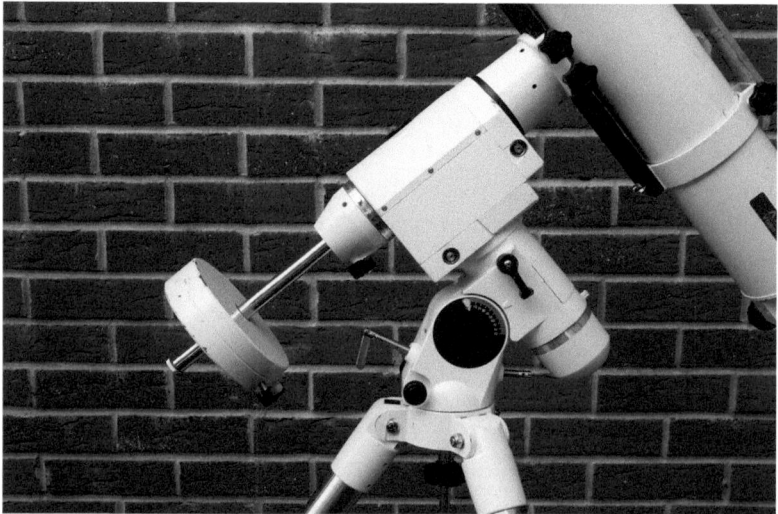

Fig. 4.1 Picture of an equatorial mount

come to a point) divided by the aperture or diameter of the lens. So, if we have a focal length of 600 mm and an aperture of 80 mm then the focal ratio will be:

$$600 \text{ mm} \div 80 \text{ mm} = 7.5$$

or a focal ratio of f 7.5. This will render a wide field of view. If the aperture is 150 mm and the focal length is 1,500 mm then 1,500 mm ÷ 150 mm = 10, so the ratio will be f10. The field of view will be smaller than the f 7.5, but the larger focal ratio has the effect of "magnifying" the image slightly and curtailing the amount of sky seen around the object. This is advantageous in some instances.

There are two main types of telescope: the refractor, which uses a lens or combination of lenses in one cell placed at the front of the telescope, and the reflector, which uses a parabolic mirror to collect the light and focus it on an eyepiece. Reflectors usually have greater aperture than refractors, they are cheaper to produce and the market is well supplied with them. Another common telescope is the mirror/lens system of the Schmidt Cassegrain and its derivatives, such as the Maksutov telescope.

The refractor uses a lens, placed at the front of the tube. In astronomical parlance this lens is known as an object glass. Light enters the lens, travels along the tube to an eyepiece placed at the focal point, which then magnifies and clarifies the image. A good quality refractor in the 100- to 120-mm range is a very versatile instrument and will provide a good platform for observing most deep sky objects. Refractors of this size, if of good quality, will provide much better images than reflectors and provide a ready platform for an SLR camera or CCD device. A good refractor is a versatile instrument, but the difficulty of making quality objective lenses in sizes larger than 150 mm for commercial sale has always provided the amateur with a problem of aperture.

To see really faint and indistinct objects, reflectors are the instrument of choice as they are very durable, portable (despite their larger size) and the sheer aperture and light grasp plays into the hands of those looking for fainter objects or more detail in the brighter ones. Reflectors therefore become the main telescope of most observers, as large apertures can be purchased for a fraction of the cost of a top quality refractor. Most reflectors are built according to the Newtonian design where a parabolic mirror at the base of the telescope reflects light back up the tube to a mirror angled at 45 degrees (a flat) and then out through the side of the tube at a comfortable height for viewing. For the price of a good quality refractor you can buy a 250-mm or 300-mm reflector on a Dobsonian mount. Despite this size the

Fig. 4.2 Refractor

instrument is still portable and can be transported easily, though it is incumbent on the observer to check that the optics remain collimated.

The main advantage of a large reflector is of course aperture. The larger apertures of commercially available reflectors offer the observer the experience of seeing fainter objects than a modest refractor for almost the same price. Many reflectors also come in short focal lengths of f4 or f5, making them rich field telescopes with a wide field of view, which are ideal for studying nebulae while the increased light grasp will pull the object out of the night sky and allow fainter details to be seen under good observing conditions. For the price, a reflector is many amateurs telescope of choice, as it combines light grasp, short focal lengths and portability married to ease of setup and use.

One difference between refractors and reflectors in practice is also that a refractor is "ready to go" soon after set up, whereas a reflector may take some time cooling down to the external temperature before it obtains fine

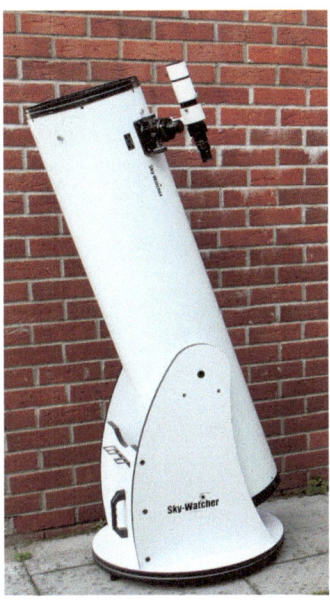

Fig. 4.3 Dobsonian reflector

images. Tube currents play a pervasive role in visual astronomy, and it is best to let a reflector settle before attempting to view any fainter objects on a target list. In addition, many of the Newtonian reflectors found for commercial sale are not built for photography but just for visual observing. Although this is not a concern for most observers, this is something to be taken into account if one uses a large reflector.

Catadioptric telescopes are almost a compromise system between reflectors and refractors in that they use both mirrors and lenses to achieve focus. Most amateurs will be conversant with the typical setup of such telescopes, known as the Schmidt Cassegrain, wherein a corrector lens at the front of the 'scope adjusts the light path to fall on a spherical mirror, which then bounces light onto the silvered spot on the lens, back down the tube and out through a hole in the primary mirror to a focus outside the rear of the mirror cell. Maksutov-Cassegrains use much the same light path, the principal difference being in the curvature of the front lens or meniscus of the system. The light path is quite long in a relatively small instrument, as the light path makes three trips around the system, resulting in a larger focal ratio, generally between f10 and f13.

Fig. 4.4 Maksutov-Cassegrain telescope

When exploring nebulae, the aperture of the telescope is of paramount consideration. The larger the aperture, the fainter in magnitude the objects that can be seen. The limits of magnitude resolution and telescope size can be worked out with a simple formula:

$$5 \log\left(D_1 / D_0\right) + 6$$

where D_1 is the telescope aperture and D_0 is the pupil of the eye (6 mm). Working out this formula gives us the following approximations:

Telescope size (mm)	50	70	100	150	200	250	300	400
Limiting magnitude	11	12	12.7	13.6	14.2	14.5	15.2	16.7

From the above it is easy to see that ANY observational aids will increase our light grasp. Telescopes have provided us with effective tools to see some of the faintest objects in the universe.

Eyepieces

With variable star observing the need for high magnification is rare; medium to low magnifications will usually suffice. In some circumstances high magnification might make the apparent field darker. This can help to identify some stars, particularly when they are a minimum brightness. Only experience will tell if it does make a difference in a particular telescope system.

A good quality eyepiece can make a huge difference to your observing experience. The eyepieces usually supplied with telescopes are quite cheap and do not bring out the best in the instrument, and the buyer should change them as quickly as possible for good quality ones to bring out the best. When buying a small telescope, do not believe the claims of the manufacturer that this instrument will magnify up to 400 or 500 times. Such claims are almost fraudulent, as at such magnifications only a blur will be observed through the eyepiece. As a general rule, a telescope is performing at its optimum when it has a magnification of ×25 per 25 mm of aperture. Therefore, if you have a 100-mm telescope, the maximum it should be permitted to magnify, with resultant clear detail, is ×100. Following this advice will forestall any frustration you will ultimately experience if you have the misfortune to be sucked in by the advertiser's misleading statements.

Like any telescope, an eyepiece will come in a choice of focal lengths. These are usually displayed on the barrel as 32 mm, 25 mm 20 mm, etc. These figures give the user the focal length of the optical elements within the eyepiece. Each eyepiece will obviously perform slightly differently on each telescope that the observer will use, and the stargazer will have to work out the magnification obtained by each eyepiece on each telescope they use (if you are lucky to own more than one!) The magnification of any eyepiece can be obtained by remembering that the focal length of the eyepiece (in mm) divided into the focal length of the telescope (again in mm) gives you the magnification with that system. So if you have a 1,000-mm focal length telescope and a 32-mm focal length eyepiece, then 1,000 mm ÷ 32 mm = × 32, or magnification of 32 times. The figure shows a variety of eyepieces.

Fig. 4.5 Picture of eyepieces

If the 20-mm eyepiece is used with a system having a longer or shorter focal length then adjustments have to be made accordingly, i.e., a 700-mm focal length will now have a magnification of ×35 (700 ÷ 20) while a 1,400-mm focal length will now have a magnification of ×70 (1,400 ÷ 20). Magnification will also impact upon the field of view of each instrument using that eyepiece, as with each successive increment the field of view shrinks and emphasizes the object under scrutiny. This can result in greater detail and contrast, but it depends upon the eyepiece and telescope in use. There are diminishing returns no matter what setup you use.

Remember, too, that any small telescope is not going to show you the wonders of the universe on the same scale as a large observatory or the Hubble Space Telescope will do. Many objects show all kinds of details, but most do not. Learn to work within the confines of what your equipment is capable of. A small telescope should be able to see many variable stars in addition to many other objects scattered around the night sky.

Small telescopes can be limiting but are very rewarding to use. Above all, observing variable stars should be calming and fun but also teach patience and perseverance. Take it only as fast or as slow, as intense or as easy, as is right for you. Remember that variable star observing is a feast and a great scientific experience, but it also has a human element to it. That human element is most present when trying to teach oneself the vagaries of variable star observation.

With the advent of digital cameras, CCD's and webcams, can variable stars be captured and their light curves followed with modern cameras? There are methods of doing this, but firstly, let us take a look at the equipment that many amateurs may already own.

Camera Equipment

The advent of digital SLR cameras and CCD imaging systems means that good quality photographs are within easy reach of most amateurs. However, adding such equipment to your personal store entails a bit of spending! Good quality DSLRs are within the budgets of most amateur astronomers, and CCD cameras can also be purchased inexpensively, though items do obviously become more expensive dependent on quality and reliability. Fig. 4.6 shows a typical series of examples: Canon 30D, Canon 400D and fitted with a T-mount and telescope adapter, a Canon 1000D.

Ideally, then, the observer should be equipped with either a CCD camera or DSLR. Choosing such equipment can be a long process of comparison and getting advice from experts in the field, but such advice is worth considering so as to avoid common mistakes. A DSLR is a versatile piece of equipment, and of course is very useful for photography outside that of astrophotography and photometry, whereas a CCD camera is not. There are many manufacturers of quality DSLRs, but the general consensus is to purchase either Nikon, Canon or Olympus cameras, as these manufacturers have a wide range of auxiliary equipment available such as lenses, T-mounts, adapters, focusing screens and filters.

Although digital, these cameras follow a similar format as 35-mm film cameras in that their sizes, weight and controls are flexible, and they are easy to set and control once you work with them for a while. It is not our intention here to recommend any particular brand, as excellent results can be had with all the above types – it is merely a choice of preference and

Fig. 4.6 A selection of DSLR Canon cameras (Image by the author.)

cost. In addition to the camera, the observer will require a cable release inimical to his or her camera system – which will cost a bit extra – to prevent shaking.

Lenses

If the observer is going to use a driven camera mount, then the choice of lens will be crucial. Occasionally a DSLR can be purchased with a choice of lenses, typically a 28- to 80-mm focal length or a larger one with 75- to 300-mm focal length, for example. These lenses are not built for astrophotography, though they do serve to provide wide-field shots of the sky. Naturally, any image gained will be very small, depending on the target subject and require a lot of enlargement and enhancement.

Lenses are obviously a feature that will require some consideration. Is the observer going to use the standard 50-mm lens that comes with most cameras or are they going to attach the camera to the telescope? If lenses are a preferred option then long focal lengths will require a driven mount as the amount of exposure will be cut down by star drift across the field.

Fig. 4.7 Assorted lenses from a 9 mm to a 250 mm

Focal Length and Field of View

Camera lenses come in a variety of sizes, and some of the most common ones for DSLR cameras have a variable focal length usually between 24 to 75 mm, 18 to 55 mm, or even long focal length ones such as 75 to 300 mm. The versatility of these lenses cannot be overstated for terrestrial photography, but there is a caveat; at night when the manual focus of these lenses is crucial, they do not always perform very well and can be difficult to focus accurately. This is due to the small amount of focal travel the lens incurs in daily operation, where the servo-motors inside the lens can move the focus quite accurately. Turn off the automatic feature, however, and the dexterity required to get the focus just right requires some patience and effort.

Usually this can be done by small movements once an infinity focus is reached. Such lenses do not have an infinity focus at the "end" of their travel, but a little way back from the end of travel. By taking several photographs in many positions at night, the observer can get the focal point as accurate as possible for that particular lens/camera combination. This author makes a small mark on the lens that is visible in red light so that faster focusing can be gained quickly.

A Telescope as a Long Focus Lens

The best approach to photographing any astronomical object is to connect the camera to a telescope at its focus point. This procedure then gives one the advantage of having a large telephoto lens with an f ratio exactly the same as the telescope. With this arrangement the observer doesn't have to do any complicated mathematics in working out f-ratios as the focal length and focal ratio of the camera and telescope system is exactly whatever is the f ratio and focal length of the telescope. One merely has to think in terms of exposure times and accuracy of guiding if necessary. With DSLRs several photographs may be taken of the one object and then stacked in the appropriate software to produce a single, higher resolution image.

CCD Cameras

There are so many variants of CCD camera available that it is not possible to cover them all in great detail and it is recommend that the observer read the reviews of these online to make the best choice. Many astronomers have

CCD cameras such as the various *Atik* cameras, models by the Santa Barbara Instruments Group (SBIG) or Starlight Express. Many of the models are full color CCDs; others are black and white and require colored filters such as RGB or BVR to produce a full color image. Once again, choice is down to observer preference and costs. The advantage of the CCD camera over the DSLR is the smaller field of view and the rapidity of capture and quality of the images, which can be manipulated in various software programs such as *Maxim DL, Artemis* and others.

Webcams specifically made for astronomy may also be employed in a similar way to CCDs, although the resolution depends on the chip. One of the most popular of such items is the DMK camera, which requires a program entitled *RegiStax* (or similar) to complete the imaging processing. If the observer wishes to use this form of photography, see *Deep Sky Video Astronomy* (Springer) by Massey and Quirk, which covers in depth the video and *Registax* techniques needed to take great pictures.

Resolution of the CCD camera or of any DSLR is the most important thing for an amateur astronomer. For the observer equipped with or contemplating purchasing a CCD camera, the chip size and pixel numbers on the axis of the chip are very important in obtaining detail and ensuring that the camera and telescope will resolve any astronomical object sufficiently.

Fig. 4.8 An ATIK 314 L CCD camera

Thankfully most astronomical bodies are quite large in comparison to the pixel size of a CCD, but to ensure that any camera you consider purchasing will be adequate for your needs use the simple equation here that will enable you to ensure that the chosen CCD camera is adequate. The relationship is:

$$R'' = \left[\left(Ps_m \, 206 \right)/a \right]/f$$

where R" is the angular resolution in arc seconds, Ps_m is the pixel size, a (in mm) is the telescope aperture and f is the focal ratio of the telescope. If the camera has pixels of 10 μ size and is fitted to a 200-mm aperture f5 telescope, then:

$$10 \times 206 = 2060$$
$$2060 \div 200 \left(a \right) = 10.3$$
$$10.3 \div f5 = 2.06$$

The chip therefore has a resolution of 2.06".

If you then need to know the size of the field of view through such a camera the simple relationship is:

$$S^\circ = \left(Pr \; Pn \right)/3,600$$

where S° is the size of the frame in degrees, Pr is the pixel resolution and Pn is the number of pixels along the axis of the chip divided by 3,600. So for a chip with 1,500 pixels along its main axis:

$$1,500 \times 2.06 = 3,090$$
$$3,090 \div 3,600 = 0.85$$

This is larger than the full Moon and will encompass most astronomical objects, but obviously smaller, more distant objects will lose some resolution despite the fact that the image can be enlarged.

No matter what the individual uses, one of the factors essential to any photography is the focus; it must be sharp and free of any obvious defects such as smearing at the edges. It may take time and practice for any astronomer to achieve a good focus, but it is well worth the effort, as anything that is out of focus is disappointing when one looks at the pictures. Focusing is a slightly frustrating task when out in the field yet is the most essential component of any photography, and getting the focus right deserves time and attention. Many DSLRs have a 'live-view' facility, a consideration to be taken into account when purchasing one, while a CCD has the ability to download a picture there and then to judge for focus and quality.

Filters for Photometry

There are filters built for the telescope, filters built for CCD imaging and even those such as those made by Hutech or Astronomik that fit the front of DSLR camera lenses and come in a variety of sizes to fit the lens optic. Generally, it does not matter who the supplier is, as most of these filters are comparable in price. It obviously depends on the kind of photography or visual work the astronomer performs as to which filter is most suitable.

To get full-color photography one must use a variety of commercially available filters. For the UBVRI filters that are commonly used with CCD cameras it is recommended that the reader refer to the following details for the typical bandwidths, filter systems and filter wheels to use with these cameras.

Considering that most CCD cameras on the market are monochromatic, it is important that any photometry is done via the V band, and so a set of BVR filters is essential. The CCD camera is going to act as the photometer, and most observations will be done in V band to gain the most accurate measurements of the stellar magnitudes across the field.

Therefore, the greatest requirement for photometry is a red, green and blue (RGB) filter system that make color rendition possible. These three colors, two of which are primary, represent the two ends and the middle of the electromagnetic spectrum of visible light. Green (or V-band) is used as the "middle" color, as the eye has a visual peak at 500 nm in the green part of the spectrum and represents a more accurate determination of true color in the spectrum. These filters, which have the collective name of photometric filters and are generally regarded as RGB or UBV, are usually joined in deep sky photography by an infrared blocking filter that is either known as an I or an L band filter, thus rendering basic CCD photography of the sky an LRGB exercise or UBVRI.

Filter sets were standardized by the work of Harold Johnson in the 1950s and then again by A. Cousins in the 1970s and are known collectively as the Johnson-Cousins UBVRI system. A 1990 paper by Michael Bessell in the *Journal of the Astronomical Society of the Pacific* outlined the filters of the Johnson-Cousins system and recommended that for budding astrophotographers, Schott optical glass filters, when placed together with the Johnson-Cousins system, work well together.

Filters were later added to these systems that incorporate the infrared wavelengths and are known as JHK filters after their band-pass in the I-R part of the spectrum, where J has a bandwidth of 1.1 to 1.4 microns, the H filter corresponds to the H infrared line at 1.5 to 1.8 microns and the K band to the infrared band 2.0 to 2.4 microns. Needless to say these filters are

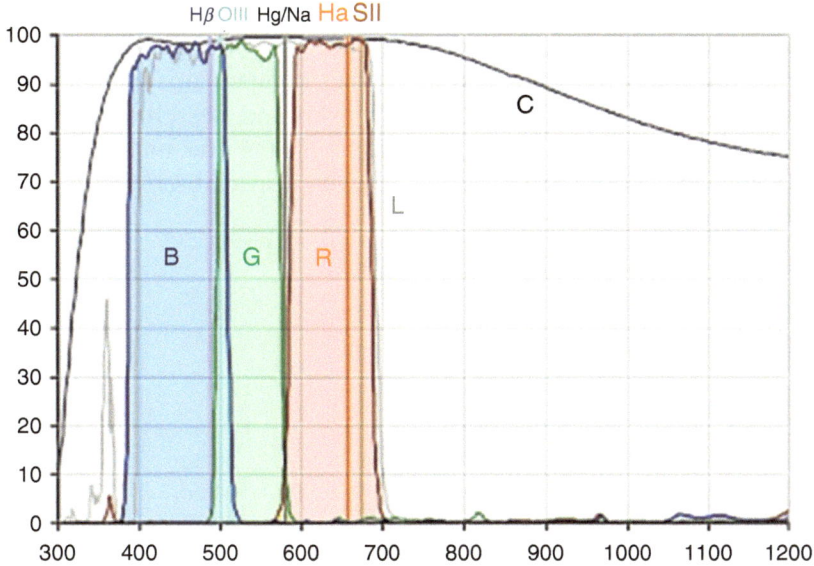

Fig. 4.9 UBVRI band-pass (Image from https://asterisk.apod.com/viewtopic.php?t= 26289.)

specialist items and will not be found among the general armory available to amateur astronomers. Additionally, filter sets have been produced for specific uses for professional observatories and are based upon photoelectric detection such as the Gunn-Griz system.

The specific bandwidths of each UBVRI filter based on Bessell imaging for astrophotography with DSLR and CCD cameras can be seen here in Fig. 4.9.

The wavelength that determines the band-pass of each filter type can be seen along the x-axis of the illustration. It is important to know that the SI unit of wavelength is the meter and that one nanometer (nm) approximates to 1×10^{-9} m. An Angstrom is 1×10^{-10} m, and although in use in astronomy, the nanometer is the preferred unit.

Each filter will therefore cover a specific bandwidth of wavelengths in color while the more narrowband filters such as OIII and Hα will allow small bandwidths centered on their particular wavelength (501 nm and 656 nm, respectively). It is the colored filters that render the image its correct appearance, and other filters tend to add to the specifics of the image or used solely to obtain detail about the object at a specified wavelength.

The common filters for UBVRI photography systems can be found in the following table. Although there are over 200 photometric systems in use, the table illustrates the most common filter components and their band-passes that correspond to the Johnson-Cousins and Bessell systems. Most amateur astronomers will utilize imaging systems that make full use of UBVRI plus the narrow bandwidths of the more specialized filters we shall discuss in due course.

The Johnson-Cousins Photometric System

Filter Letter	Band-pass	Full Width Half Maximum
U	365 nm	66 nm
B	445 nm	94 nm
V	551 nm	88 nm
G	550 nm	90 nm
R	658 nm	138 nm
I	806 nm	149 nm
Z	900 nm	125 nm
Y	1,020 nm	120 nm
J	1,220 nm	213 nm
H	1,630 nm	307 nm
K	2,190 nm	390 nm
L	3,450 nm	472 nm
M	4,750 nm	460 nm

Using filters with a monochrome CCD camera to obtain a true color image follows on from the kind of photography that we will discuss in a moment. Modern photography with DSLR cameras incorporate colored filters on the chip, known as the Bayer system, to render true color. The filter pattern on the chip is generally 50% green, 25% red and 25% blue, thus gaining the most band-pass from the visible, but the filtration systems that can be added to the lenses will involve either neutral density filters or the more special-ized Hα. Beyond this DSLR cameras or CCD cameras will be preferred by the observer.

CCD chips are generally very sensitive to red, and infrared and modern filters are constructed of special Schott glass to accommodate this, which does differ slightly from the bandwidths of the Bessell system noted above, but not so much that this system is completely changed. As an example, the I band filter in the Bessell system is now generally referred to as the Ic and has a steep cut off at 900 nm with no light transmission beyond this. It is also no longer made of Schott glass and has an additional dielectric coating to block near infrared. This is valuable for CCD chips that are sensitive to these long wavelengths, especially if studies of very red stars are a part of the observing program. However, for most photographers the minutiae of detail will not be that important. All the user wants is for the filter to work at the appropriate wavelength.

CCDs are the industry standard for photometry, but increasingly there is interest in using standard digital SLR cameras with good quality lenses and stable, driven mounts. We shall examine using CCD's and DSLR cameras for photometry in a later chapter.

Chapter 5

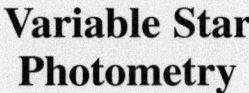

Variable Star Photometry

There are many classes of variable star, and each class represents a different way that a star can vary. Intrinsic variable stars will change in size, shape or temperature over time, or they may experience changes in light due to physical processes around the star such as shells or disks. Others undergo eclipses by binary partners. There are a huge variety of such stars, and not all of them are going to be easy targets for amateurs.

When you look up at the night sky on a clear night, each star looks like a distinct point of light. However, Earth's atmosphere smears this point into a tiny disc and the worse the atmospheric conditions are, the bigger this disc becomes. The differences in disc size is due to atmospheric seeing, and to ensure that we want to use all the light coming from the star in our magnitude estimates, we need to gain all this light and measure it as accurately as possible.

All the information a star can provide us with comes from the light we receive. Interpreting the light and the message it contains is a bit like a detective story. We have all the clues from our images, but we need to piece the information together to ensure that we are going to get the right information that leads to our magnitude estimates and anything else we wish to learn about the star. The amount of radiation in the form of light we receive from a star is known as the flux. The science of measuring this flux is known as photometry, and among other things, this information not only gives us the magnitude of the source but can also help in us understanding other parameters such as star temperature, size, mass, luminosity, and

© Springer Nature Switzerland AG 2018
M. Griffiths, *Observer's Guide to Variable Stars*, The Patrick Moore
Practical Astronomy Series, https://doi.org/10.1007/978-3-030-00904-5_5

chemical composition and provide us with a basic understanding of the physics of stars. The information also can be used to calculate the distance to an object using the tools in our chapter on astrophysics, as one can see, Photometry is an extremely versatile tool.

Starting in Photometry – Some Tools of the Trade

To perform photometry, all the observer needs is to have access to a CCD camera and a computer. It is also possible to perform basic photometry with DSLR cameras, but we shall come to that subject presently. CCD detectors and their computerized software enable us to do this with a high degree of accuracy. Both The BAA and the AAVSO have online programs and guides that will give the amateur all the information necessary to perform photometry of variable stars. This chapter will provide a quick guide. In addition to the information here, there are several books on CCD photometry that will be very useful, such as *A Practical Guide to Light Curve Photometry* by Brian D Warner; *An Introduction to Astronomical Photometry* by E. Budding and O. Demirican; and *Astronomical Photoelectric Photometry* by F. B. Wood.

CCD cameras operate in monochrome, and color images are obtained by passing the incoming light through a filter system. The most obvious filters that will be required are the red, green and blue ones that make color rendition in cameras possible. These three colors, two of which are primary, represent the two ends and the middle of the electromagnetic spectrum of visible light. Green (or v-band) is used as the "middle" color, as the eye has a visual peak at 500 nm in the green part of the spectrum and represents a more accurate determination of true color in the spectrum. In variable star photometry, the green or "V band" is the most appropriate one to use, as it mimics the eye in its interpretation of brightness.

These filters, which have the collective name of photometric filters are generally regarded as RGB or UBV and are usually joined in deep sky photography by an infrared blocking filter that is either known as an I or an L band filter, thus rendering basic CCD photography of the sky an LRGB exercise or UBVRI. The specific bandwidths of each UBVRI filter based on Bessell imaging for astrophotography with DSLR and CCD cameras can be seen here in Fig. 5.1.

Since you most likely will be spending more time working with your data at the computer than actually taking images at the telescope, it is important that you have some basic computer skills. You should also understand the software you are using very thoroughly – not only how to use it but the

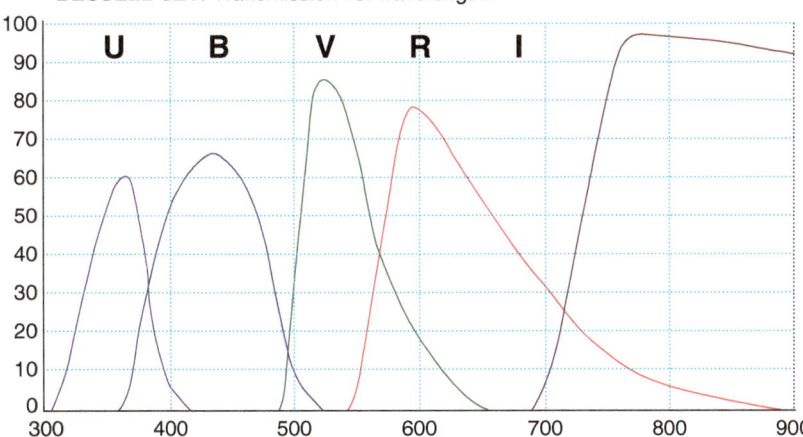

Fig. 5.1 UBVRI band-pass (Image from http://www.aip.de/en/research/facilities/stella/instruments/data/johnson-ubvri-filter-curves.)

basics of what it does. Taking some time to learn how to use your software correctly will quickly pay off.

As you will soon discover, it doesn't take very long for you to begin accruing lots of images that will consume a lot of storage space on your computer. You should decide how you will handle this in an organized manner before you start. Many observers save their data to large external hard drives, as their computer hard drive sometimes is not large enough to keep all the images necessary after a few observing runs. Everyone makes mistakes or misses problems with images, and it is not uncommon for observers to find a processing error or a comparison star sequence change. Therefore, it is essential that your files are complete and organized so you can find what you need as easily as possible.

As a basic step in organizing yourself, the amateur can remember the following. Keep nightly logs containing notes on what is being observed, weather, the Moon phase, etc. Keep a file of your calibration images; in addition, keep the nightly raw images you take and the calibrated images, which include the flat field and dark subtracted images that you would ordinarily perform for CCD imaging. Lastly, keep a written log of what you are doing. Ideas on such logs were given in a previous chapter.

As with image acquisition and calibration, there is software available to do most of the hard work, but it is important that you understand it and use it properly, or your results may not be scientifically useful.

Software and Data reduction

There are also several websites and guides on which software to use to reduce your data through Microsoft Excel and easy to use programs such as Makali'i, which is an *MS Windows*-based package that allows the user to get the data from the FITS files taken by the CCD; this is available at https://makalii.mtk.nao.ac.jp.

Alternatives such as the Aperture Photometry Tool can be found at http://www.aperturephotometry.org. or Photometry Express at http://www.photometryexpress.co.uk by the late John Moore at Fareham Astronomical Society. There are many others, and the observer can choose the one they are most comfortable with. For the purposes of this chapter, we will use Makali'i as an example. There are several excellent guides to using this software available, via the Gaia website at https://www.gaia.ac.uk/sites/default/files/resources/Photometry_with_Makali.pdf, which was written by the staff at Gaia and the Faulkes Telescopes (LCOGT). There is also a guide from the NAO site mentioned above. In addition, there is some software available from the AAVSO that will enable you to plot light curves from the data collected by the AAVSO, so if you want to use these tools, then take a look at what is on offer at https://www.aavso.org/vstar or in their software directory at: https://www.aavso.org/software-directory, including the excellent *Pleione* software that generates a finder chart for a given variable star (like planetarium software), is able to show the AAVSO observing chart with comparison stars, allows the recording of observations and generates reports in an AAVSO acceptable format.

The most important thing to do in variable star research is of course to plot the star's light curve. Some types of objects, such as binary stars will demonstrate a repeat pattern in these measurements indicative of two objects orbiting a common center of mass, while others will produce sinusoidal light curves indicative of intrinsic activity. Light curves vary due to the nature of the stellar systems and it is not always possible to plot a "simple" light curve with some of the variables that will feature in this book.

Resolution, Image Size and Field of View

Your camera and telescope work together to define the resolution and field of view (FOV) you can expect from your system. It is important to quantify these and design an observing program that takes advantage of the strengths of your setup. You will also need to understand the importance of the image intensity against background noise and the background sky.

When performing photometry, a value for the aperture has to be selected so that stellar intensity can be accurately judged. If this value is underestimated, measurements will not include all the light being collected by the telescope. Conversely, if it is overestimated, it is possible that too much background light from nearby stars not related to the variable star might be included. In any astronomical image, the photons of light gathered from the source will follow a distribution of a particular shape known as a Gaussian or bell-shaped curve. From this curve, a figure known as the full width half maximum (FWHM) can then be calculated, which gives us an optimum aperture size to select for our analysis. The curve can be seen in Fig. 5.2.

The full width half maximum (FWHM) is therefore a measure of the quality of an astronomical image based on how much the telescope and the atmosphere have smeared a point source over several pixels in the CCD. When you inspect the image of a star, you will notice that it is made up of a group of pixels, with some brighter ones near the center and some dimmer ones surrounding it. Ideal images of point sources made by optics have an intensity pattern called an Airy disc. However, in practice starlight has to pass through Earth's atmosphere that diffuses and expands the pattern. In order to measure the intensity of an image like this when it doesn't have sharp edges, the term full width half maximum (FWHM) is used. This can be defined as the number of pixels that are filled to one half

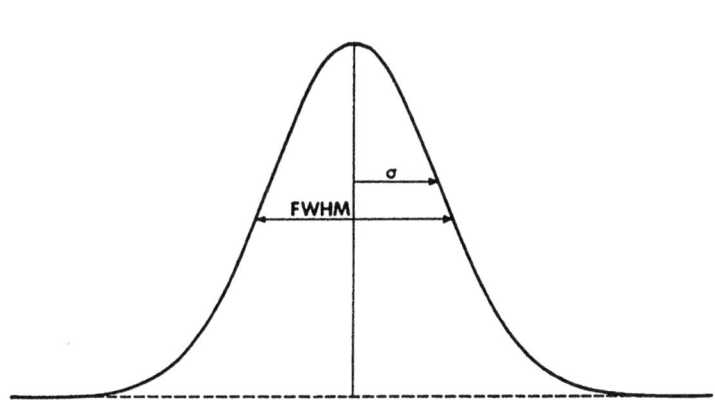

Fig. 5.2 Gaussian curve (Image from https://www.itl.nist.gov/div898/handbook/pmc/section5/pmc51.htm.)

of the range between the background and the brightest pixel in the star's image.

Another piece of information it would be useful to know about your system is image scale or resolution. The image scale of your system can be computed using this equation:

$$\text{Image scale} = (\text{CCD pixel size / focal length}) \times 206.265$$

(Image scale in *arcsec* / pixel, CCD pixel size in microns, focal length in millimeters.)

You should be able to get the CCD pixel size from the manufacturer's specifications on your camera. The focal length of your telescope can also be expressed as f/ratio times the aperture.

Knowing the image scale of your system is handy for figuring out how the seeing conditions are in your location on any given night. Simply use this equation:

$$\text{Seeing} = \text{Image scale} \times \text{FWHM}$$

One can use the transparency and Bortle scales already discussed in this book to estimate the conditions during your observing run. On average, good seeing allows the astronomer to separate objects up to 2 arc seconds apart with a good CCD camera and telescope, although the stellar image will be subject to such vagaries as seeing, angular resolution of the telescope and the resolution of the CCD chip.

The images you create with your CCD camera will be saved in the FITS file format. FITS (flexible image transport system) is the standard method of storing scientific images into computer-readable files and is supported by all software packages. A useful feature of the FITS format is that information about the image, such as target name, time of exposure, etc., can be stored in a readable format along with the image itself.

As we want to measure the change in brightness of an object with time, we need many sets of images that combine an accurate record of those brightness changes and how the photon count between them all differs so that a light curve can be drawn.

Photometry Basics

There are two kinds of photometry that are commonly done in astronomy, differential photometry and all-sky photometry. As observers using this book are determining the magnitudes of stars, the photometry we shall concentrate on is differential or aperture photometry. In this way the observer can derive magnitudes for a variable star that can be compared to the known magnitudes of stars in the field of view.

Since the effects of the atmosphere and the size of our telescope determine the size that each star appears on our image, we need to take account of this before we can measure the stars properly. Once we have imaged our star field with the variable star in it, we can use the computer software to place a circle over the star in question and measure its intensity. We can consider this to be the number of photons collected by the telescope during the exposure, and our software should be able to make a determination of what the magnitude of the star should be.

To do this, one must calculate a value for the radius over which we wish to measure our circular star. Once a radius value is chosen based on the FWHM, it is important to use the same radius for all images for a particular dataset. This should always be the case if you are doing photometry of a variable star in order to create a light curve over time.

However there are some caveats, depending on the software used. The observer can find an optimum aperture for the equipment by choosing several values for the aperture radius (say between 5 and 40), and plotting a graph of radius against number of photon counts. As you increase the aperture radius, you can see that the count rate also increases. This is because more of the star is included within the aperture. Beyond a certain value, however, the curve flattens, showing that the star is already and you are now just adding the background sky. It is worth noting, though, that there is not necessarily one unique correct answer here; rather, any aperture within a couple of pixels will probably work well.

Here are a few other suggestions regarding the size of the aperture rings in your software package. Ensure that the diameter of the star aperture is 3 to 4 times the rough average FWHM of all the stars you wish to measure. Make sure that the brightest star you are planning to measure fits within the star aperture you have selected. If the aperture is too small it won't measure the star completely. If the aperture is too large, you increase the chance of inadvertently including other faint stars in it. Remember that as a general rule the diameter of the inner circle of the sky annulus should be about 5 times the average FWHM (or about 10 pixels across). As with any photometry of variable stars, remember to adjust the outer ring of the sky annulus if necessary. A bigger sky annulus can get a better signal to noise ratio, but there is a rule of diminishing returns. Once you have calibrated the best FWHM and aperture ring, it will generally do for most of your observations.

Once you have some images of the field where the variable star can be found, it is good to ensure that your check stars, the stars you are using for variable comparison, are not variables themselves! A check star is simply a star of known brightness that doesn't vary that can be treated in the same way as you treat your target star. You should be able to compare the magnitude you determine for it with its published magnitude, and the results should be very close. Ensure as far as you can that your check star is as similar in color and magnitude to the variable as possible.

What makes the image useful for variable star astronomy is that the image is also tagged in some way in the FITS file with the time it was taken. So at this point you have everything you need, namely a measurement of light at a specific moment in time. However, this is just the first step. Beware also of errors in your data. To get the best results you should sample the images so that the FWHM of your seeing disk is spread across two or three pixels or can be as accurate as possible. This will help to optimize the signal-to-noise ratio of the chip and improve the accuracy of your measurements.

Often, achieving this sort of goal accuracy is very difficult, given that it is dependent on the seeing conditions and limitations of your equipment. Nevertheless, one may be able to tweak it to gain the best conditions possible. If you are averaging a FWHM of less than 2 pixels, you are probably under sampling the image. If the FWHM of your seeing disk is more than 3 pixels in diameter, you may be over sampling the image. Fortunately, there are things you can do to remedy the situation.

Binning is something you can do to increase your effective pixel size by grouping pixels together. Your software can be set up to sample (or bin) a group of 2 pixels by 2 pixels to make those 4 pixels act as one. There is a tradeoff, however. Resolution will be lost, so you have to be sure that star images have not blurred together with other nearby stars. Also if one of the four pixels in the group is saturated, the accuracy of the photometry will suffer. Your calibration frames must also be binned to the same degree. It is not really recommended to bin groups larger than 2×2 pixels.

In most modern software, such as the ones discussed earlier in this chapter, a click of your mouse on the stellar target will provide you with the magnitude of your object. You are now ready to use the images and software to produce a light curve using *MS Excel* or other appropriate packages. As a quick guide the following mantra can be used and memorized:

1. Check your images.
2. Identify the stars, including your variable.
3. Set the aperture.

4. Choose the check and comparison stars carefully.
5. Measure the magnitudes.
6. Determine the uncertainty.

Using Makali'i

Makali'i, or Subaru, is a free, *Windows*-based piece of software that allows the user to view FITS images and to perform aperture photometry upon them. Its unusual name derives from the fact that it is the Hawaiian name for the Pleiades star cluster, which is also known as Subaru in Japanese. The URL for the software is as above, and excellent guides to its use can be found at the National Schools Observatory website at https://www.schoolsobservatory.org/discover/projects/clusters/makalii and at the Japanese National Astronomical Observatory website https://makalii.mtk.nao.ac.jp.

Once we have obtained an image of our variable star via our CCD camera, the resultant image (in the form of a FITS file) shows several dots or small circles representing the individual stars. The circle's size is partly determined by the brightness of that particular star. In turn, the areas of bright starlight on the dark background are determined by the number of photons that hit the camera's detector during the exposure. From this point on, the observer can use a photometry package such as Makali'i to "count" these photons to give information about the brightness of the objects in the image.

In Makali'i, aperture photometry can be done by means of a single click for each star, and the results are then displayed in a window that can also easily be saved in a format that is usable in packages such as *MS Excel*. In addition to photometry, Makali'i can also be used for blinking, stacking and aligning images, combining of color images, spectroscopy and for calibrating images using flat fields, dark and bias frames. Images can also be saved in a variety of formats including JPEGS, bitmaps and PNG files. The opening screen of the package looks like Fig. 5.3, and as one can see, it is very user friendly.

Once you have opened the software click the *Open* button in the toolbar to display the file opening dialog. Select a FITS file, and you will see the image will be displayed in a new window. You can move a mouse over this window to show the x and y coordinates in the information bar towards the top left of your screen. Additional information such as photon count, the value of each individual pixel and the average count value of the overall

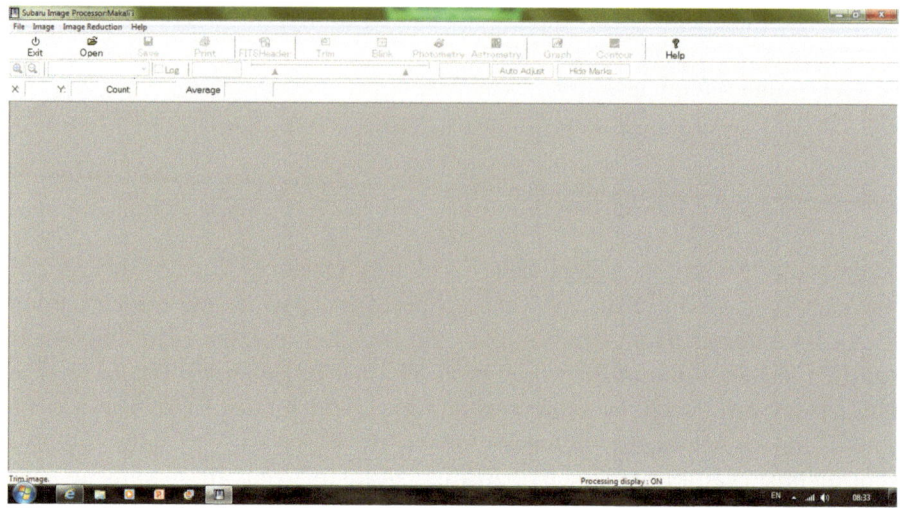

Fig. 5.3 Opening screen of Makali'i (Screen shot by the author.)

image will also be displayed here. If the FITS header of the image contains WCS (World Coordinate System) information, then the RA and Dec of your objects can be ascertained as a check that you have the correct field. Using the other buttons and menus in this screen will enable you to alter the contrast of the image, rotate the image and convert to a gray scale image where the stars are black dots on a white background.

Aperture photometry using this software allows a measurement of the brightness of an object to be made. To start this process, click the *Photometry* icon in the top toolbar or use the dropdown menu that can be found under *Processing* and then choose the *Aperture* mode. The *Aperture Photometry* box that opens has some options that you will need to change.

For example, the indicator *Radius* will appear, and this should be changed for best results to *Semi-Auto*. This feature then performs aperture photometry based on the star with the star, the sky and the sky radius specified by the observer. In addition, the *Find Center* box should be ticked, as this corrects for slightly inaccurate placing of the cursor.

The observer is now ready to measure each star, and this is done by simply clicking on each star. You will notice that each star clicked will have a set of values that are added to the *Aperture Photometry* box. From this data you will obtain your magnitudes and light curves.

The *Aperture Photometry* box collects a lot of information and values in a layout that may not be as easy to digest as other programs you may be familiar with. However, using the *print* option for these values, a file can be

Number	Obj x	Obj y	Obj Radius	Obj Pxl	Count	Mean	Obj SD	Sky X	Sky Y	Sky InRadius	Sky Width	Sky Pxl	Sky Count	Sky Mean	Sky SD	Result count	Radius
1	254	987	6	149	20,000	139.77	139.77	254	987	10	3	150	1700	15.89	4.9999	19889	SEMI
2	239	976	6	150	20,000	140	140	239	976	10	3	150	1700	15.887	4.9989	19877	SEMI
3	230	970	6	150	20000	141.77	141.77	230	970	10	3	150	1699	15.778	4.8899	19776	SEMI
4	267	992	6	151	20000	141.99	141.99	257	992	10	3	150	1701	15.889	4.9987	19867	SEMI
5	278	994	6	149	20000	141.98	141.98	278	994	10	3	150	1699	15.778	4.8899	19978	SEMI
6	264	968	6	149	20000	141.88	41.88	264	968	10	3	150	1701	15.889	4.9988	19986	SEMI

Fig. 5.4 Excel document from *Aperture Photometry* (Screen shot by the author.)

Number	Obj X	Obj Y	V Intensity	V Mag	B Intensity	B mag

Fig. 5.5 Makali'i magnitude estimates (Screen shot by the author.)

created that will open in *MS Excel* that places the photometry values into columns, as can be seen in Fig. 5.4.

As you can see, the files that Makali'i creates have several columns in them that show a wealth of data, including the brightness of the sky around the stars along with some values that the software creates for clarity.

A – Number. Ideally, this will match the star numbers in the finder chart.
B – Obj X. This is the x coordinate for your star.
C – Obj Y. This is Y coordinate for your star.
Q – Result Count. This is the number of counts for a given star.

If you have been able to successfully perform photometry on the same number of stars the numbers that you have in columns B and C for your two files will look very similar. The only reason these columns might not be identical is that the telescope may have drifted slightly between the two exposures. It is always worth checking that these figures correspond. Once you are sure that everything is in order, you can then calculate the magnitudes for your check stars and the variable star in question.

For most variables it should be possible to collect values for around 5 to 10 stars, each of which will then have an intensity value in the V band. You can of course use the B and R band, too, but it's the mid-range V band that is most important to variable star observers. The data gathered here can now be converted to magnitudes using the following formula:

$$\text{Magnitude} = -2.5^* \, \log \left(\text{counts} \, / \, \text{exposure time} \right)$$

To get the best results it is best to convert from the above spreadsheet by placing the values one needs into a separate worksheet. The worksheet should look like Fig. 5.5.

From the last spreadsheet you can paste columns A, B, C and Q into columns A to D of your new worksheet. You can then select column Q from your B-band.csv file and paste it into column F of your new sheet. Columns E, G and H can remain empty.

Next, you can calculate values for the V magnitude column (column E) by typing the formula:

$$= 2.5 \times \log(D/5) \text{ where D is the corresponding value in column D.}$$

This can be repeated for the B magnitude value into column G using

$$= 2.5 \times \log(F/5) \text{ where F is the corresponding value in column F.}$$

Finally, column H is calculated by subtracting column E from column G. Once this is done you should have a set of magnitudes for all the stars in the field, including your variable star. That can then either be entered into your observing log or notebook or recorded in some other fashion.

Once you have several sets of such data you can amalgamate your magnitudes to make a light curve of the variable star by using either the graph feature in *MS Excel* or in some fashion that you are familiar with. The AAVSO and the BAA provide good guides in order for the observer to reduce their magnitude readings into an acceptable light curve.

DSLR Photometry

One of the great things about modern DSLR cameras is that the CMOS or CCD sensors in DSLR cameras are very similar to the sensors in astronomical CCD cameras. If you are going to use a DSLR camera, you have one advantage as a photometer over CCD photometry. The R, G and B (BVR) images are recorded simultaneously, thereby reducing acquisition time. However, there are a few issues with one-shot color imaging that have significant implications for photometry. Additionally, interchangeable lenses provide an additional advantage, allowing different configurations depending on the needs of the individual observing program. As a caveat, fixed focal length lenses are preferred, as they tend to have fewer air-to-glass surfaces and lower distortions compared with zoom lenses.

In the Bayer filter array used in the chips of DSLR cameras, half the pixels record green light and one quarter of the pixels record red or blue light. Software that interprets and displays these images has to estimate how much red, green and blue light would have fallen on each pixel in the image.

It does this by looking at the mid-range surrounding the green pixels and interpolating how much green light should have fallen on the red and blue pixels. Therefore, an average is taken and then computed for the green range or V band, and this mid band is the most useful for photometric work, as it mimics the visual. Once these features are recognized, atmospheric extinction can be applied if necessary to your final estimate of the star's magnitude.

Instead of focusing accurately, a DSLR has the advantage that any defocused star image actually covers more of the chip. More individual pixels of each color are illuminated, allowing better photometry. Furthermore, longer exposures are possible before saturation occurs, so total flux, or light gathered, is increased.

Defocusing extends the star image over many more pixels, which leads to better color interpolation. It also allows longer exposures before pixel saturation, hence better counting statistics. DSLR cameras also have long exposure and noise reduction programs built in, although many who use DSLR cameras in photometry recommend not using these as they reduce contrast and wipe out much of the image obtained by the defocused image. Choose a moderate ISO and open the camera shutter to f2.8 or f4 if you are using a lens and gauge your results. Remember that every camera will be slightly different.

In addition, DSLR cameras offer a variety of file formats once an image is taken. Put the file format to RAW, as this records directly what the sensor has detected and includes no processing or compression by the camera. The file extension used by Canon for RAW files is .CR2, while Nikon use . NEF. Olympus and Pentax use RAW images, too, but consult the manual if your camera is from another manufacturer. RAW requires an enormous amount of memory storage, so a large card would be very useful, as every bit of information is necessary for accurate photometry.

Obviously most DSLR cameras used for photometry will be attached to a telescope to narrow the field of view. It is therefore important that the same telescope and f ratio is used with the same settings on the camera at all times so that you are taking images at an absolute standard as far as one possibly can. For additional reading on this matter I would refer the reader to the excellent article in the Journal of the BAA by Des Loughney, which is available here:

http://www.britastro.org/vss/JBAA%20120-3%20%20Loughney.pdf

Or the article from the Journal of the AAVSO by Brian Kloppenborg et al that was published in 2012 and is available here:

https://arxiv.org/pdf/1303.6870.pdf

Such cameras and telescopes may be a bonus for long period variable observation since if the same exposures and same settings are used every time a photograph of an object is taken good comparisons can be made with stars of known magnitude in the field. This may well alleviate the need to use one as a photometer, but results should be carefully compared to existing magnitudes and light curves.

The raw images obtained by your DSLR can then be examined in such software as *Maxim DL* or similar and aperture photometry applied in a similar manner to the foregoing. The AAVSO also provides a wonderful guide to DSLR variable star observing and also the software to reduce your DSLR results into a spreadsheet that enables one to judge the magnitudes correctly. It is highly recommended that the observer get these guides and digest them to gain the best from their observations.

The AAVSO has an absolutely wonderful guide to DSLR photometry written by experts with years of experience. Anyone contemplating DSLR photometry should download this guide, which can be found at https://www.aavso.org/sites/default/files/publications_files/dslr_manual/AAVSO_DSLR_Observing_Manual_V1.4.pdf, where all the information they will need can be found in a user friendly format. The BAA variable star section also has such a guide, and together these will prove invaluable and go beyond anything that can be covered here. It's always a good idea to use the experts! The BAA guide can be found at https://www.britastro.org/vss/DSLR_PHOTOMETRY.pdf.

Obviously the foregoing is only a brief guide and is not meant to be a full-on introduction to photometry. We have not covered taking the same number of CCD images in B or R band, which can also be done, neither have we mentioned the importance of dark and flat frames, which are very important in all CCD imaging and will be necessary for DSLR imaging also. The observer using such cameras will no doubt be familiar with these datasets and how to obtain them, and we leave a fuller description of their uses to others with more experience. Nevertheless, we hope that the foregoing will be useful to the observer.

Chapter 6

Observing Variable Stars with Binoculars

A good pair of binoculars is one of the most underrated pieces of equipment available to any observer. It may seem strange that variable stars can be observed in such equipment, but a good pair of binoculars will reveal much more than a poor telescope. It cannot be stressed sufficiently; invest in a good pair of binoculars rather than a cheap telescope. Binoculars are monetarily cheaper than a telescope. Even better, they can be used in the day as well as the night, whereas a telescope for astronomical use does not have this advantage.

Compared to a telescope, binoculars do have certain advantages. Although they are smaller and have lower magnifications, they are lighter and easier to take outside and use. They give a much wider view than a telescope, thus making objects easier to find. Binoculars also let you use both eyes, providing surer, more natural views. In addition, everything seen through them is rendered the correct way up, not the upside-down or backward the way that a telescope presents.

It is not possible to recommend the best types of binoculars, as optics do vary between different manufacturers, but Olympus, Leica, Minolta, Zenith and Swift all make beautiful optics with excellent quality, though as a general guide it is essential to remember this: The larger the aperture, generally, the clearer the view, but the less magnification the better. This can be demonstrated by what is known as the aperture index of binoculars.

Binoculars have two numbers set into them near the eyepieces or quoted on the box. These are usually rendered as 10×50 or 10×25 or any

© Springer Nature Switzerland AG 2018

M. Griffiths, *Observer's Guide to Variable Stars*, The Patrick Moore Practical Astronomy Series, https://doi.org/10.1007/978-3-030-00904-5_6

combination of numbers. What this means for a pair of 10 × 50 binoculars is that they have a magnifying power of ×10, and the aperture of the main object glass is 50 mm. To ascertain the aperture index, simply divide the magnification into the aperture, which with a pair of 10 × 50s is:

$$50 \div 10 = 5$$

A pair of 8 × 40s will have the same index, and a pair of 7 × 50s will have an index of 7. The larger the aperture index, the better for stargazing they will be. An index of around 4+ is good; anything below that is not really useful. It is the field of view that is all important, and as the light path of Porro prism binoculars may differ in comparison to Roof prism ones, such differing paths may result in a smaller field of view.

Our advice when purchasing any binoculars would be to try out as wide range as possible until you find one that satisfies your requirements, feels comfortable and one with which you can actually find objects with little fuss. Eye relief is also a consideration, especially if you wear eyeglasses, and a binocular with good eye relief allowing a long light path beyond the magnification lenses is ideal. Pictured here is a selection of binoculars ranging from 20 × 80 giant binoculars through the standard 10 × 50 on the left, 7 × 35 in the center (which have wonderful eye relief) and the 8 × 42 on the right, again with very good eye relief and adjustable lens caps.

Fig. 6.1 Picture of binoculars

Obviously this kind of selection will range in price, but it cannot be stressed enough that a good pair of binoculars are an excellent tool for looking for some variable stars. The magnitude ranges of some variables stretch over just one or two magnitudes, and when one is contemplating a variable with a range between magnitude 4 to 8, then binoculars will prove to be sufficient as long as the observer has patience and proficiency.

One drawback that is a constant refrain at star parties – how do I keep the binoculars still enough to observe something in the sky? Many people complain that they cannot see too much through binoculars as they cannot keep them from moving or stay dead on target due to the movements of the body. Actually you can turn this to your advantage. If you have an astronomical object in your sights, gently move the binoculars in a rounded motion around the target, and the eye will naturally stay on the target regardless of what you are doing. Similarly, with sighting the object initially, just look at the area of sky with your target in it, and without moving your head simply bring the binoculars up to your eyes. You should have little difficulty in seeing the magnified image at once. Try both of these methods if you do have difficulty looking for and maintaining a lock on objects. Of course in today's market you also have image stabilized binoculars, which prevent shaking of the image no matter what the body movement is (unless you fall over), but such items can be very expensive!

If you are observing with a pair of binoculars, especially if they are the giant binoculars that can be obtained today, you will find after a while that your arms are beginning to tire and that you have developed a pain in the middle of your back caused by bending over backwards to bring the instrument to bear on stars near the zenith. To deal with both these problems it is advantageous to invest in a tripod to steady the binoculars, preferably of a type that has a canted head, enabling an observer to get under the tripod to observe the zenith. Even an ordinary tripod is a step in the right direction, as a properly mounted pair of binoculars will show much more detail in hazy objects than an unsteady hand-held pair of even the finest binoculars. Of course, with giant binoculars, mounting them properly is a must, as these instruments are virtually wide field telescopes and must be treated as such.

For any observations of variable stars, the binoculars are best supported by a tripod.

For binocular observing, as with any other form of variable star observation, it is important to know which types of variable we are dealing with. Apart from some variables, which defy being 'boxed,' there are four major classification groups, described as eruptive, pulsating, cataclysmic and eclipsing. The subtypes are many, and differing forms of variation may be occurring in a single star or a stellar system. The main classes and subtypes of variable suggested for binocular use are:

Eruptive variables, e. g., GCAS, RCB
Pulsating, e. g., M, SR, RV,
Cataclysmic, e. g., N, NR, ZAND
Eclipsing, e. g., subtypes EA, EB, EW

The *General Catalogue of Variable Stars* (GCVS) is the best source for descriptions of the categories. Dependent on an individual's interests the longer period objects may be more appropriate for observers with limited time to spend searching. Weather conditions and seasonal timing may also play a part in these decisions. Observers with a great deal of spare time may wish to concentrate on eclipsing binaries that eclipse in one night, as these stars require an estimation of the magnitude every 10–15 min around the time when an eclipse is predicted. Depending on the variable star, the timings of such eclipses can be found in the annals of the AAVSO or the BAA. The basic visual observation in all these cases is an estimate of the star's magnitude as compared with 'standard stars' at that time. How this is achieved is as follows.

Making and Recording Your Observations

To the inexperienced observer, the prospect of measuring the magnitude of a star to a few tenths of a magnitude may seem daunting, but after some practice it is surprisingly straightforward. The most important piece of equipment initially is an accurate variable star chart, as the magnitudes obtained by commercial planetarium software can sometimes lead to inconsistent results when the data is analyzed and compared with official charts. Charts are available from the AAVSO or the BAA.

The variable star in question is then compared in brightness with usually at least two of the comparison stars with known fixed brightness, one brighter and the other a little fainter than the variable. The Pogson step method is the recommended analysis, as it requires the eye to recognize differences of one tenth of a magnitude. The fractional method simply requires the brightness of the variable to be expressed as a fraction of the difference between two comparison star magnitudes.

However, there are several hurdles to overcome before a confident observation can be made. Correctly identifying the variable itself is the most common problem faced and also the Purkinje effect, which affects the eye's peak sensitivity, shifting it towards the blue end of the spectrum at low light levels. This has consequences for recording red variables.

Variable star observers should have the object in view for as long as possible and are then required to round the number to a tenth of a magnitude. Often there is no point in reducing the observation to a final magnitude while out in the field, since this may be done indoors. If there are any other items that may have affected the observation, such as like the Moon being nearby, haze/thin cloud or light pollution, then note these details.

It is also important to note observer bias. Not everyone has the same level of expertise or an appreciation for the finer points of magnitude differentiation. In addition, the nature of the star being observed is an important factor to take into consideration when you attempt to put together an observing program. The slow variation of Mira stars, for example, need not be observed more than three times a month – perfect if you can only manage the odd night here and there to observe. Eruptive and cataclysmic stars, however, need to be monitored on a nightly basis, which takes a little more dedication. Bias is a major problem to inexperienced and experienced observers alike, and it's important not to over observe a variable. A star that varies over a period of 400 days will change in brightness very slowly indeed, so any change in its magnitude is not likely to be noticed on consecutive nights. However, the memory of the observation made a night earlier will still be fresh in your mind and will be difficult to ignore. Plan your observing sessions around the period of the stars you are observing.

Once you feel comfortable and confident enough after some practice then begin recording your star (for your own records that can be transferred to the AAVSO light curve data) as $A - 2, C + 3$ with the convention that minus means 0.2 is added (fainter than) to A's value. The plus 3 means 0.3 is subtracted (brighter than) from C's value. The variable's magnitude is deduced as $[(7.0 + 0.2) + (7.6 - 0.3)]/2 = 7.25$, and this is rounded to 7.3.

The list below is based on lists from the AAVSO and gives any variable star observer a wide selection of objects to concentrate on. Observers can pick as many types as one likes as part of their program or concentrate on just one type. Finder charts can be obtained from the AAVSO, and their variable star plotter is an invaluable tool that can be found at https://www.aavso.org/apps/vsp/.

To make the observation, compare the brightness of the variable to two or more of the comparison stars on the chart – one fainter and one brighter. As mentioned earlier, two methods are in common use, the fractional and the Pogson step method, the former being more suited to beginners. The Pogson step is recommended for those who have some experience in this type of observing, as it requires the eye to recognize differences of one tenth of a magnitude. The fractional method simply requires the brightness of the variable to be expressed as a fraction of the difference between two comparison star magnitudes.

In all instances of variable star observation, it is very important to allow time for your eyes to become adapted to the dark. This allows you to see fainter stars. Additionally, the color sensitivity of our eyes change as they dark adapt, and it is important for them to be fully dark adapted so that one can compare stars in a consistent way every time they are observed.

Which then brings us to recording red variable stars. The sensitivity of the eye to red light varies from person to person. However, it is quite common for an individual's brightness estimates of red stars to differ from those made by other observers by several tenths of a magnitude. Another problem observers encounter is when you stare at a red star, it will appear to brighten compared with other stars! This will obviously affect your brightness estimate. Hence staring at stars should be avoided. Short glances will produce a more accurate estimate.

In addition, always observe when the star is well above the horizon –at culmination if at all possible. Stars that are closer to the horizon will appear to be fainter, as they have a greater depth of atmosphere for the light to travel through. Nor should you wander all over the sky looking for comparison stars; always use comparison stars that are nearly at the same altitude as the variable star.

When using binoculars or a telescope, always bring the variable and comparison star in turn to the center of the field of view. If a star cannot be seen by direct vision, then it may be glimpsed by using averted vision. Always record when the variable was glimpsed with averted vision, as this affects the limiting magnitude that you can observe.

Remember, too, that there will be nights when the atmosphere is so unsteady that it is impossible to make accurate estimates. Stars may fade or brighten relative to each other as you watch them. There is little that you can do in such circumstances other than to see if conditions have improved after an hour or more or to try again on another night.

The following list gives a fairly comprehensive overview of over 100 binocular variables visible in both the northern and southern hemispheres. Enjoy viewing and recording these fairly bright variable stars.

Table 6.1 List of binocular variables

Name	N/S	RA & Dec	Const	Type	Period	Mag
EG And	North	00 44 37.19	And	ZAND+E	482.57	6.97–7.8 V
V0370 and	North	01 58 44.33	And	SRB	119	6.85–8.05 V
tet Aps	South	14 05 20.10	Aps	SRB	111	4.65–6.20 V
R Aqr	North	23 43 49.46	Aqr	M + ZAND	387	5.2–12.4 V
R Aql	South	19 06 22.25	Aql	M	270.5	5.5–12 V
UU Aur	North	06 36 32.84	Aur	SRB	235	5.10–6.6 V
psi 1 Aur	North	06 24 53.90	Aur	LC	--	4.54–5.7 V
R Boo	North	14 37 11.58	Boo	M	223.4	6.2–13.1 V
RV Boo	North	14 39 15.86	Boo	SRB	228	7.5–8.7 V
RW Boo	North	14 41 13.38	Boo	SRB	209	7.5–8.6 V
RX Boo	North	14 24 11.63	Boo	SRB	158	7.0–8.3 V
X Cnc	North	08 55 22.88	Cnc	SRB	180	5.69–6.94 V
RS Cnc	North	09 10 38.80	Cnc	SRB	242.2	5.33–6.94 V
RT Cnc	North	08 58 16.00	Cnc	SRB	90.04	7.05–8.1 V
V CVn	North	13 19 27.77	CVn	SRA	191.89	6.52–8.56 V
Y CVn	North	12 45 07.83	CVn	SRB	267.8	4.86–5.88 V
TU CVn	North	12 54 56.52	CVn	SRB	44.2	5.50–6.2 V
W CMa	South	07 08 03.44	CMa	SR	160	6.27–7.09 V
VY CMa	South	07 22 58.33	CMa	LC	--	6.5–9.6 V
RT Cap	South	20 17 06.53	Cap	SRB	422	6.8–8.0 V
R Car	South	09 32 14.60	Car	M	307	3.9–10.5 V
S Car	South	10 09 21.89	Car	M	149.49	4.5–9.9 V
U Car	South	10 57 48.19	Car	DCEP	38.829	5.74–6.96 V
AG Car	South	10 56 11.58	Car	SDOR	371.4	5.7–8.3 V
BO Car	South	10 45 50.66	Car	SRC	130.7	7.15–7.94 V
BZ Car	South	10 54 06.25	Car	SRC	111.3	6.86–8.6 V
CK Car	South	10 24 25.36	Car	SRC	266	6.85–8.06 V
EV Car	South	10 20 21.60	Car	SRC	825	6.76–8.8 V
HR Car	South	10 22 53.84	Car	SDOR	--	6.8–8.80 V
ASAS J110135– 6102.9	South	11 01 35.76	Car	LC	--	6.73–7.91 V
V0465 Cas	North	01 18 13.88	Cas	SRB	60	6.1–7.2 V
rho Cas	North	23 54 23.03	Cas	SDOR:	--	4.1–6.2 V
R Cen	South	14 16 34.32	Cen	M	502	5.3–11.8 V
T Cen	South	13 41 45.56	Cen	RVA	181.4	5.56–8.44 V
RV Cen	South	13 37 36.05	Cen	M	457	7.0–10.8 V
V0744 Cen	South	13 39 59.81	Cen	SRB	90	4.83–6.95 V
V0766 Cen	South	13 47 10.86	Cen	EB + SDOR:	1304	6.11–7.50 V
V0854 Cen	South	14 34 49.41	Cen	RCB	--	6.84–15.11

(continued)

Table 6.1 (continued)

Name	N/S	RA & Dec	Const	Type	Period	Mag
W Cep	North	22 36 27.56	Cep	SRC	350	7.02–8.5 V
SS Cep	North	03 49 30.02	Cep	SRB	90	6.5–7.7 V
miu Cep	North	21 43 30.50	Cep	SRC	835	3.43–5.1 V
T Cet	South	00 21 46.27	Cet	SRB	159.3	4.96–6.90 V
omi Cet	South	02 19 20.79	Cet	M	331.96	2–10.1 V
tet Cir	South	14 56 43.99	Cir	GCAS	--	4.81–5.65 V
R CrB	North	15 48 34.41	CrB	RCB	--	5.71–15.2 V
RR CrB	North	15 41 26.23	CrB	SRB	60.8	7.3–8.2 V
SV Crv	South	12 49 47.03	Crv	SRB	65.2	6.75–7.6 V
RU Crt	South	11 51 06.47	Crt	SRB	60.85	7.43–8.59 V
BH Cru	South	12 16 16.79	Cru	M	521	6.55–10.1 V
W Cyg	North	21 36 02.50	Cyg	SRB	131.7	5.10–6.83 V
RS Cyg	North	20 13 23.66	Cyg	SRA	417.39	6.5–9.5 V
AF Cyg	North	19 30 12.85	Cyg	SRB	92.5	6.4–7.7 V
CH Cyg	North	19 24 33.07	Cyg	ZAND+SR	--	5.6–10.1 V
V1070 Cyg	North	21 22 48.60	Cyg	SRB	62.4	6.56–7.56 V
U Del	North	20 45 28.24	Del	SRB	120	6.14–7.61 V
CT Del	North	20 29 26.35	Del	SRB	83.5	6.8–8.5 V
EU Del	North	20 37 54.71	Del	SRB	58.63	5.41–6.72 V
R Dor	South	04 36 45.60	Dor	SRB	172	4.78–6.32 V
AY Dor	South	06 31 01.11	Dor	SR	86.4	6.93–7.94 V
UX Dra	North	19 21 35.52	Dra	SRB	175	5.94–7.1 V
AH Dra	North	16 48 16.63	Dra	SRB	158	6.4–8.6 V
Z Eri	North	02 47 55.92	Eri	SRB	74	6.17–7.18 V
RR Eri	North	02 52 14.19	Eri	SRB	94.6	6.80–7.62 V
BM Eri	North	04 13 29.62	Eri	SR	559	6.76–8.03 V
BR Eri	South	03 48 47.53	Eri	SRB	73.3:	6.5–8.16 V
TV Gem	North	06 11 51.41	Gem	LC	--	6.27–7.5 V
pi 1 Gru	North	22 22 44.18	Gru	SRB	195.5	5.31–7.01 V
X Her	North	16 02 39.17	Her	SRB	102	5.8–7.0 V
ST Her	North	15 50 46.63	Her	SRB	144	6.8–8.3 V
UW Her	North	17 14 24.54	Her	SRB	103.6	6.8–8.7 V
AC Her	North	18 30 16.24	Her	RVA	75.29	6.85–9.0 V
IQ Her	North	18 17 54.84	Her	SRB	76.5	6.8–8.0 V
OP Her	North	17 56 48.53	Her	SRB	120.5	5.85–6.73 V
V0939 Her	North	17 10 18.53	Her	LB	--	7.47–8.50 V
g Her	North	16 28 38.55	Her	SRB	89.2	4.3–5.5 V
R Hya	North	13 29 42.78	Hya	M	380	3.5–10.9 V
U Hya	North	10 37 33.27	Hya	SRB	183.1	4.56–5.4 V
V Hya	South	10 51 37.25	Hya	SRA	530.7	6.0–12.3 V

(continued)

Table 6.1 (continued)

Name	N/S	RA & Dec	Const	Type	Period	Mag
Y Hya	South	09 51 03.72	Hya	SRB	154.4	6.2–7.4 V
RT Hya	North	08 29 41.15	Hya	SRB	245.5	7.0–10.2 V
RV Hya	South	08 39 43.76	Hya	SRB	108.7	7.31–8.58 V
CL Hyi	South	02 27 46.84	Hyi	SRB	75	6.71–7.75 V
T Ind	North	21 20 09.48	Ind	SRB	272	5.76–6.47 V
R Leo	North	09 47 33.49	Leo	M	309.95	4.4–11.3 V
R Lep	North	04 59 36.35	Lep	M	445	5.5–11.7 V
S Lep	North	06 05 45.54	Lep	SRB	97.3	6–7.58 V
RX Lep	North	05 11 22.85	Lep	SRB	79.54	5.12–6.65 V
Y Lyn	North	07 28 11.61	Lyn	SRC	110	6.58–8.25 V
CE Lyn	North	07 44 09.49	Lyn	SRB	--	7.3–8.02 V
XY Lyr	North	18 38 06.48	Lyr	SRC	120	5.6–6.6 V
HK Lyr	North	18 42 50.00	Lyr	SR	186	7.5–8.4 V
T Mic	North	20 27 55.19	Mic	SRB	352	6.74–8.11 V
CF Mic	South	21 25 28.36	Mic	SRB	56.05	7.1–8.3 V
U Mon	South	07 30 47.47	Mon	RVB	91.32	5.45–7.67 V
X Mon	North	06 57 11.81	Mon	SRA	155.8	6.8–10.2 V
RV Mon	North	06 58 21.49	Mon	SRB	121.3	6.88–7.7 V
SX Mon	North	06 51 57.33	Mon	SRB	77.67	7.43–8.21 V
BO Mus	South	12 34 54.46	Mus	SRB	132.4	5.3–6.56 V
X Oph	North	18 38 21.13	Oph	M	338	5.9–8.6 V
V0533 Oph	South	17 53 03.32	Oph	SR	69.7	7.06–8.24 V
W Ori	North	05 05 23.72	Ori	SRB	212	5.5–6.9 V
BL Ori	North	06 25 28.18	Ori	SRB	153.8	5.9–6.6 V
BQ Ori	North	05 57 07.39	Ori	SRB	243	7.1–9.0 V
S Pav	South	19 55 13.96	Pav	SRA	390	6.6–10.4 V
Y Pav	South	21 24 16.73	Pav	SRB	449	5.79–6.90 V
GO Peg	North	22 55 00.97	Peg	SRB	79.3	7.14–7.91 V
SU Per	North	02 22 06.89	Per	SRC	533	7.2–8.7 V
CI Phe	South	23 44 19.19	Phe	SRB	739	7.08–8.20 V
R Pic	South	04 46 09.55	Pic	SR	168	6.35–10.1 V
Z Psc	North	01 16 05.03	Psc	SRB	155.8	6.37–7.49 V
TV Psc	North	00 28 02.84	Psc	SR	49.1	4.65–5.42 V
L2 Pup	North	07 13 32.32	Pup	SRB	140.6	2.6–8.0 V
WX Ret	South	03 37 44.56	Ret	SRA	266	7.78–9.52 V
RY Sgr	South	19 16 32.77	Sgr	RCB	37.67	5.8–14.0 V
UX Sgr	South	18 54 54.48	Sgr	SRB	95.7	7.25–8.3 V
V1943 Sgr	South	20 06 55.24	Sgr	SRB	330	7.1–8.4 V
AH Sco	South	17 11 17.02	Sco	SRC	735	6.7–9.5 V

(continued)

Table 6.1 (continued)

Name	N/S	RA & Dec	Const	Type	Period	Mag
BM Sco	North	17 40 58.55	Sco	L	--	5.25–6.46 V
SW Scl	North	00 06 14.20	Scl	SRC	143.6	7.3–9.32 V
R Sct	South	18 47 28.95	Sct	RVA	146.5	4.2–8.6 V
tau 4 Ser	North	15 36 28.19	Ser	SRB	86.7	5.89–7.07 V
Y Tau	North	05 45 39.41	Tau	SRB	245	6.4–7.3 V
RX Tel	South	19 06 58.21	Tel	LC	--	6.45–7.47 V
BL Tel	South	19 06 38.11	Tel	EA/GS+SRD	778	7.09–9.08 V
W Tri	North	02 41 30.57	Tri	SR	108	7.4–8.4 V
X TrA	South	15 14 19.17	TrA	SR	361.1	5.03–6.05 V
DM Tuc	South	22 57 05.78	Tuc	SRB	74.58	6.88–8.19 V
Z UMa	North	11 56 30.22	UMa	SRB	195.5	6.2–9.4 V
RY UMa	North	12 20 27.33	UMa	SRA	310	6.49–7.94 V
ST UMa	North	11 27 50.38	UMa	SRB	80.88	6.0–7.2 V
TV UMa	North	11 45 35.04	UMa	SRB	53.74	6.72–7.45 V
VW UMa	North	10 59 01.80	UMa	SRB	615	6.69–7.71 V
V UMi	North	13 38 41.07	UMi	SRB	73	7.06–8.7 V
GO Vel	South	08 37 39.69	Vel	SRB	520	6.56–7.80 V
RT Vir	North	13 02 37.98	Vir	SRB	157.9	7.41–9.0 V
RW Vir	South	12 07 14.90	Vir	SRB	175	6.71–7.50 V
SS Vir	North	12 25 14.40	Vir	SRA	361	6.0–9.6 V
SW Vir	North	13 14 04.38	Vir	SRB	146	6.2–8.0 V
BK Vir	North	12 30 21.01	Vir	SRB	140	7.28–8.8 V
FP Vir	North	13 35 52.06	Vir	SRB	67.8	6.55–7.69 V
FI Vul	North	20 48 51.18	Vul	SR	--	6.97–8.30 V

Chapter 7

Giant Stars and Their Variability

Long-period variables (LPVs) and pulsating variables (PV) are comprised of a huge variety of stars that show great promise to any observer who wishes to take up the discipline of recording variable starlight curves. Their longer periods, and in the case of the red giant stars, large range of magnitudes, enable the observer to record over an extended period of time; the interference from inclement weather is reduced, as missing data is not as important in long term periodicity as it would be with cataclysmic variables. Additionally, such stars are visible for long stretches of the year and only out of observational range for a few weeks of the year, when in conjunction with the Sun. Therefore, these stars form a very important backbone of variable star observing.

Long-period variables are not really a specific type of variable star but are a large and diverse group of different types that are collected under this rubric. Strangely, the *General Catalogue of Variable Stars* does not define a long-period variable star type, although it does accept Mira-type stars as long-period variables. The term LPV was first used in the 19th century, before more precise classifications of variable stars were created, in order to differentiate groups that were known to vary over periods of 100 days or more. Studies of such stars in the 20th century determined that long-period variables were generally cool giant stars, but as the group was so large and so many stars could be included, the term LPV was restricted to the coolest pulsating stars, almost all of which were Mira-type red giant variables.

© Springer Nature Switzerland AG 2018   111
M. Griffiths, *Observer's Guide to Variable Stars*, The Patrick Moore
Practical Astronomy Series, https://doi.org/10.1007/978-3-030-00904-5_7

Nevertheless, the AAVSO traditionally define such stars as having periods from a few tens to 1,000 days, and under this term, Cepheids, RR Lyrae stars, and delta Scuti variables are LPV pulsators, but a caveat here – their periods are too short to be included as long-period variables. The simple reason for this definition is that all stars of intrinsic variable type pulsate, so definitions have to be included that ensure that the observer is aware of the evolutionary state of the star and its potential age. Additionally, stars that were previously classified as long-period variables can be seen to be distinct populations of stars, even though their underlying physical mechanisms may be similar. It is generally held that stars with variable periods of less than 200 days belong to the Population II stars of the galactic nucleus and halo, while those of longer periods are typically Population I-type stars of the galactic disk. However, such hard and fast rules do not always apply.

Therefore, long-period variables make up quite a few kinds of pulsating stars in terms of spectral type, stellar mass, and stage of evolution. Classification for some of these stars is difficult, either because the star in question has not been well observed or some physical characteristic of the star is distinctly different from a common type. Variable star observers are familiar with the AAVSO pages that describe LPVs in a section on their web page known as the "LPV Zoo," which can be found by following this link: https://www.aavso.org/aavso-long-period-variable-section.

Many long-period variable stars are the red giant types that follow the class of Mira-type variables, named after the prototype star omicron Ceti, or Mira (which translates as "The Wonderful"). This star was known to a number of ancient astronomers as a variable star and was commented on by David Fabricius and others in medieval times. Mira-type long-period variables are generally low mass stars that are undergoing the asymptotic red giant phase of their evolution. Studies by GALEX and other satellites have revealed the mass loss that accompanies such stars and the misshapen outline they have as strong convection currents distort the usual spherical shape of stars, as can be seen in Fig. 7.1.

Generally, their light curves remain fairly constant for many years, and the only differences they regularly show are slight changes in minimum and maximum magnitudes. Mira itself can reach a minimum of 14th magnitude and a maximum of 1st magnitude, but the changes generally range between 10th and 3rd magnitude. The evolution of such stars at these late phases of life are very interesting, and several subtypes such as irregular and semi-regular LPVs are known, such as μCephei or even α Orionis. As such stars have been studied for over a century, the wealth of data accumulated is an important adjunct to determining the underlying processes within such stars.

Fig. 7.1 Mira in visible and UV light from GALEX (Image from http://www.galex. caltech.edu/media/glx2007-04r_img02.html.)

Pulsating Variables

ACEP

These we have seen before, as they are classed as anomalous Cepheids. These stars show periods characteristic of variables of 0.4 to 2 days, but are considerably brighter in magnitude. ACEP stars are accepted to be metal-poor A and early F-type stars.

ACYG

These stars are variables of the α Cygni (Deneb) type, which are non-radially pulsating supergiants of Bep-AepIa spectral types and are giant stars on the main sequence rather than old evolved types. The light changes are small, with amplitudes of the order of 0.1 magnitude, and are irregular, with cycles from just days to weeks being observed among the class.

BCEP

BCEP stars are variables of the β Cephei type, pulsating stars of spectral class O8 to B6 stars with light and radial velocity variations caused by radiative pressure and gravitation capture. Variable period ranges are small, just 0.1 to 0.6 days, and their light amplitudes range from 0.01 to 0.3 magnitudes in visible light. Interestingly, these stars have a maximum brightness corresponding to its minimum stellar radius, so the radiation output is more intense as the star shrinks.

BXCIR

BX Circinus stars are hydrogen-deficient B spectral types, which reveal low-amplitude variations in light (only 0.1 mag. in visible light) and radial pulsations driven by the κ (kappa) mechanism. They show a unique and very regular period of around 2 to 3 hours on average, but the light amplitude is quite small. Studies of these stars would benefit from photometry rather than visual observations.

CEP

Cepheid variables of classic type. We have come across these pulsating stars before, and as a recap, Cepheids are radially pulsating, high luminosity variables with periods in the range of 1 to 135 days and amplitudes from several hundredths of a magnitude up to 2 magnitudes in the visible range. There are several subtypes, but it must be remembered that some DCEP and CW stars (see below) are quite often called Cepheids because it is difficult to discriminate between them on the basis of their light curves over 3 to 10 days.

CW

Variables of the W Virginis type. These are pulsating variables of the galactic spherical component and are therefore Population II-type stars with periods of approximately 0.8 to 35 days and amplitudes from 0.3 to 1.2 magnitudes in the visible band. These stars obey a period-luminosity relation that is different from that for δ Cep variables (classical Cepheids). In fact, for an equivalent period the W Virginis variables are fainter than the δ Cep stars by up to 0.7 to 2 magnitudes. This shows them to be lower mass stars, and this is borne out by observations of them in globular clusters and at high galactic latitudes. They may be separated into the subtypes CWA and CWB, where CWA-type W Virginis variables have periods longer than 8 days, while the CWB-type W Virginis variables have periods shorter than 8 days. Occasionally, the observer will come across a subset of these stars, which are named after their precursor type, BL Herculis variables.

DCEP

These are the classical Cepheids, or δ Cephei-type variables, as we have explored in several sections of this book. They are placed here as examples of a pulsating variable star, but the DCEP stars are of course young objects that have left the main sequence and evolved into the instability strip of the Hertzsprung-Russell diagram. Here, they obey the well-known Cepheid period-luminosity relation and belong to the young disk population (Pop I). Such stars have also been observed in open clusters. There are a few subsets of the DCEP type; DCEP(B) are Cepheids displaying the presence of two or more simultaneously operating pulsation modes and vary over a range from 2 to 7 days. Additionally, there is the DCEPS class, the stars of which have light amplitudes of less than 0.5 of a magnitude and almost symmetrical light curves with periods that do not exceed 7 days.

DSCT

The well-known δ Scuti-type variable star is a pulsating variable of spectral types A0 to F5 III-V, displaying light amplitudes from 0.003 to 0.9 mag. in the visible range and thus indicate a set of stars that are mostly evolving off the main sequence. The shapes of the light curves, periods and amplitudes usually vary greatly. Radial as well as non-radial pulsations are observed.

The variability of some members of this type appears sporadic and sometimes completely ceases, so they are deserving of study. There are several subsets, such as the DSCTC stars, a low-amplitude group of δ Scuti variables with a range less than 0.1 magnitudes in visible light. These are joined by HADS, or high amplitude δ Scuti stars, which are radial pulsators showing asymmetric light curves with steep ascending branches but small light amplitudes of around 0.1 magnitude.

GDOR

As we have encountered before, these are γ Doradus stars, which are non-radial pulsators, generally recognized as dwarfs with luminosity classes IV and V and revealing spectral types A7 to F7 showing one or multiple frequencies of variability. Light amplitudes do not exceed 0.1 magnitudes, and their periods usually range from just 0.3 to 3 days.

RR

Just like DCEP above these variables show similar behavior with the exception that the periods are shorter and the light output does not show as great a range. RR Lyrae types are radially pulsating giant A or F spectral type stars that have light amplitudes from 0.2 to 2 magnitudes and periods that last from 0.5 to 4 days in general. Also known as cluster variables, since they appear in great numbers in old globular clusters, RR Lyrae stars are Population II stars that belong to the spherical component of the galaxy. These variables are pulsating horizontal-branch stars, like the classical Cepheids, but their smaller mass sites them at the bottom of the instability strip, and their light curves differ significantly as can be seen from Fig. 7.2. Their light curves are very broadly similar to the classical Cepheids, indicating the κ mechanism at work, but within the type there are some significant differences in light curves. Because of this they have been recognized to have three additional sub types: RRB, RRC and RRD.

RRAB

These are RR Lyrae variables with asymmetric light curves revealing steep ascending branches and slower declines, with periods ranging from 0.3 to 1.2 days and amplitudes from 0.5 to 2 magnitudes. The RRC variables are

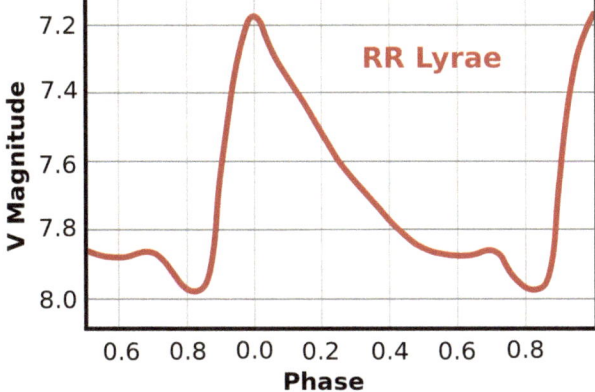

Fig. 7.2 Cepheid light curve

different in that their light curves show sinusoidal curves with periods from 0.2 to 0.5 days and amplitudes not greater than 0.8 magnitude. The stars must therefore be quite small and are still evolving along the horizontal branch. The RRD stars are similar in period and variation, but spectral research has revealed that they vary over 0.5 days with a small magnitude range (0.3 magnitudes).

SPB

These are main sequence stars of B spectral type and are massive blue-white stars with a small amplitude range of 0.2 magnitudes over periods that range from 0.4 to 5 days.

Irregular Variable Stars

Slow irregular variables are stars that show very little evidence of any periodicity, or to put it finely, it is difficult to define any variable period for them, as they have no symmetry to their light curves. Many of these stars have the label "L," as such variables have poorly defined periodicity and brightness changes. It is even possible that some of these classes may not be variable stars at all! The L types include red giants and supergiants, indicative of the swelling of their distended outer atmospheres over longer periods of time than the smaller radial pulsators.

We shall deal with these in due course, but it may be instructive to see the types of semi-regular variables drawn together under the SR rubric. SR stars, as the letters suggest, are semi-regular variables, usually giants or supergiants of intermediate and late spectral types showing a large periodicity in their light outputs and light curves. Occasionally, these stars show strange light changes that reflect the behavior of their outer swollen atmospheres. Semi-regular variables and most distended red giant stars have periods that lie in the range from 20 to 2,000 days or more. Each star has a slightly different light curve, as no one star is truly alike, and the motions of their atmospheres, although having the same underlying process, happen in a rather chaotic fashion. In semi-regular variables, the light curve reveals a small change in magnitude, usually between 1 and 2 magnitudes in the visible range. These stars can be further defined into a number of subgroups, as we can see here. There are classes SR – A, B, C and D.

SRA

These semi-regular variable stars of late-type spectral classes M, C, S or Me are red giants displaying periodicity and usually small light amplitudes ranging over 2 magnitudes. Their light curves reveal periods in the range of 35 to 1,200 days. Most of these stars almost follow the classical Mira profile, with the exception of their smaller light amplitudes. The red giant Betelgeuse is a typical semi-regular star of this type, and its irregular light curve can be examined here in Fig. 7.3.

SRB

SRB stars are again late-type classes of M, C, S or Me red giants with poorly defined periods with cycles between 20 and 2,000 days, but also showing alternating intervals of slow irregular changes and constant light curves. No one is quite sure what is occurring here, but one suggestion is that the stars may be part of a binary star system that is coalescing. SRC-type variables are very similar, with amplitudes of about 1 magnitude and periods from 30 days to several thousand days and absolute magnitudes in the range of −5 or −6. A typical example of this class is the red giant in Cepheus μ Cephei, a distended red giant also known as Herschel's Garnet Star.

Alpha Orionis (Semiregular)
1911-2001 (10-day means)

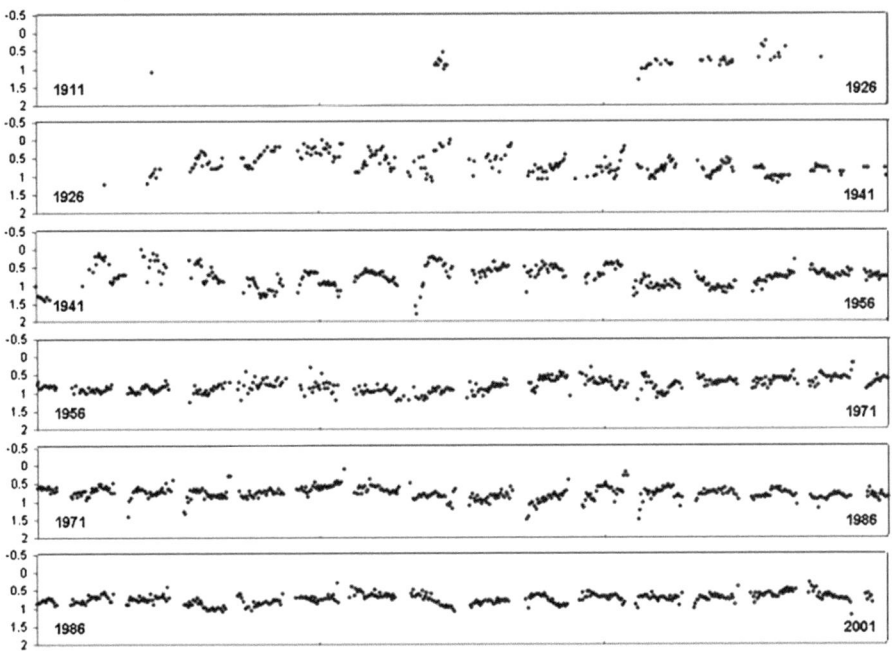

Fig. 7.3 Betelgeuse light curve (Image from https://www.aavso.org/vsots_ alphaori.)

SRD

Differing from the other semi-regular variables are the SRD stars that demonstrate semi-regular variability but are stellar giants and supergiants of F, G, or K spectral types. They also have the classes Fe, Ge, and Ke to denote the emission lines in their spectra. This group has a larger light output range with amplitudes between 01 and 4 magnitudes and variable periods from 30 to 1,200 days. If you are to observe such variables, then it would be wise for you to pick one with a larger amplitude!

Other examples of semi-regular variables are noted below. As you can see, such semi-regularity is due to the late stages of stellar evolution, and they cover almost the full range of spectral groups of high mass stars.

PVTEL

These are irregular variables named after the precursor of their class, PV Telescopii. These stars are hydrogen-deficient supergiants but not red giants. They are generally quite hard to follow, as the light amplitudes are small and the variation of period within the type is narrow and irregular. They are divided into three main subgroups of which PVTELI is the first, characterized by hydrogen-deficient A or late B spectral-type supergiants showing low-amplitude quasi-periodic variations that are probably due to radial pulsations in their atmospheres over the course of 5 to 30 days in total. Spectra have shown the radial pulsations behind such variability.

PVTELII

The second of these classes are hotter, larger hydrogen-deficient O or early B-type supergiant stars that vary over a period of 5 to 10 days but with low amplitude light curves revealing a range of only 0.2 magnitudes at best.

PVTELIII

The third PVTEL class is yet another hydrogen-deficient variable, but in this case carbon-rich spectra have been obtained that almost make such stars seem to resemble the R Coronae Borealis types. These stars are F or G supergiants showing low amplitude light variations on a timescale of 20 to 100 days but without the disappearing act of RCB stars. They remain variant over a 0.2 magnitude range with periods of between 5 and 20 days.

The Big Pulsators – Red Giants

Red giant stars are the end product of the evolution of low and medium mass stars. As such, they remain on the red giant branch for many millions of years, and as their cores (in lower mass stars at least) turn from radiative to convective motions, the star can become distended and misshaped, as many radial overtones occur at the same time. A typical example of this behavior is the variable star Mira, which has been imaged by Hubble and GALEX to be revealed as an odd, un-spherical, oblate shape. R Cassiopeia has also been shown to be similar in its behavior.

These stars have the variable appellation "L," but not all members of this class are confined to red giant status. Indicative of the evolution of red giants to the horizontal branch, some are K, G and F types, too.

LB

The classical LB stars are generally defined as slow irregular variables of late spectral types such as K, M, C and S, and as such, they are orange or red giants. The LB class can vary by several magnitudes (between 3 and 9 magnitudes), but their variable periods are not well defined. The LC types undergo irregular variability even though they are supergiants of late spectral types and have small light amplitudes in the range of about 1 magnitude.

M

The M-type variables are the classical Mira types. These stars are long-period giants with characteristic late-type emission spectra (Me, Ce, Se) and light amplitudes that range from 2.5 to 11 magnitudes in the visible range and slightly less bolometrically. Their periodicity is well pronounced in their light curves, with typical sharp rises and falls, but their periods range between 80 and 1,000 days. If we could see them in infrared, we would find rather disappointing stars with a range of less than a magnitude in IR. The light curve of o Ceti is shown here in Fig. 7.4 as an example of the type of long-period light curve we would expect to record.

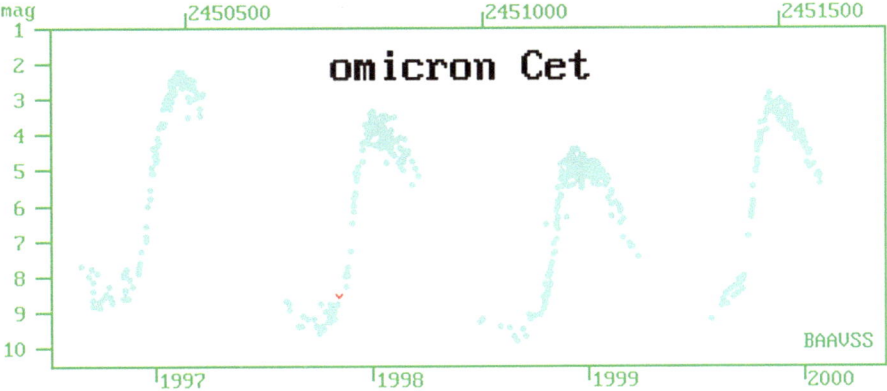

Fig. 7.4 Mira light curve (Image from http://www.britastro.org/vss/.)

There are between 6,000 to 7,000 known stars of this class, and all are red giants that have atmospheric surfaces that pulsate in such a way as to increase and decrease in brightness over periods ranging from about 80 to more than 1,000 days.

In the case of Mira, the star is on the asymptotic giant branch, and its increases in brightness make the star shine at magnitude 3.5 generally, which makes it stand out among the fainter stars of the head and neck of Cetus. However, the individual cycles can vary, and the star can go as high as magnitude 2 in brightness and as low as 4.9 magnitude, resulting in a brightness range at maximum over 15 times the luminosity at bright maxima than bright minima here. Mira has a minimum range that does not vary as much as its maximum, as records suggest, at most a magnitude range of between 8.6 and 10.1 magnitudes. These variations take place over a range of 300 days or so with a climb to maximum taking about 100 days and the decline to 9th magnitude (its usual range) taking a little over 200 days on average.

Mira is a beautiful star to watch and record, as its long period and wide range of magnitudes enable a good light curve to be obtained even when data may be missing from solar conjunction or bad weather. Again, the star is distended, and the Hubble Space Telescope has shown this wonderfully misshapen star as can be seen in Fig. 7.5.

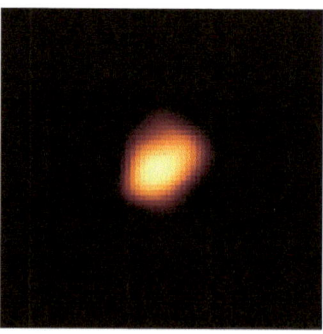

Fig. 7.5 HST Mira (Image from https://www.skyandtelescope.com/observing/mira-makes-january-nights-wonderful/.)

RV

Another class of long period variables that have a fair amplitude are the RV Tauri type. These stars are radially pulsating supergiants having spectral types F to G at maximum light and K to M at minimum. The light curves are complex, and observations reveal that they have alternating primary and secondary minima that change place over time, and so primary becomes secondary and secondary primary before the whole oscillation mode reverses to come back to "normal" again. RV Tauri types generally show a range of light output between 3 to 4 magnitudes, and such stars have a period of between 30 and 150 days.

There are two subtypes of this class but neither of them are easy objects to observe. The RVA variables that do not vary in mean magnitude but oscillate about a small undulating light curve, while the RVB types have periods of between 600 and 1,500 days but vary over a mean of 2 magnitudes. Changes in the spectra of these RV Tauri variables can be rapid and dramatic. Some of these stars show such significant changes over a few weeks that they appear to be completely different objects. The spectral changes in nearly all of these stars are poorly studied, and the connection to their light changes is generally unknown.

For those observers who are looking forward to the challenge of long-period variable stars, the following list may give one some food for thought and form the basis of a good observational program for northern and southern hemisphere observers that will keep one engaged for many years to come.

Table 7.1 Long-period variables

Long Period Variable List

Name	Constellation	Max magnitude	Min Magnitude	Range	Period	Type
R And	Andromeda	5m.8	14m.9	9.1	409 d	Mira variable (M)
S And (Supernova 1885)	Andromeda	5m.8	16m	10.2		Supernova (SN1)
U Ant	Antlia	8m.1 (p)	9m.7 (p)	1.6		LB
θ Aps	Apus	6m.4 (p)	8m.6 (p)	2.2	119 d	Semiregular (SRB)
η Aql	Aquila	3m.48	4m.39	0.91	7.1 d	Classical Cepheid (DCEP)
R Aql	Aquila	5m.5	12m.0	6.5	284 d	Mira variable (M)
V Aql	Aquila	6m.6	8m.4	1.8	353 d	Semiregular (SRB)
R Aqr	Aquarius	5m.8	12m.4	6.9	387 d	Mira variable (M)
T Aqr	Aquarius	7m.2	14m.2	7	202 d	Mira variable (M)
U Ara	Ara	7m.7	14m.1	6.4	225 d	Mira variable (M)
R Ari	Aries	7m.4	13m.7	6.3	187 d	Mira variable (M)
U Ari	Aries	7m.2	15m.2	8	371 d	Mira variable (M)
ε Aur	Auriga	2m.92	3m.83	0.91	27.08 years	Eclipsing binary Algol type (EA/GS)
R Aur	Auriga	6m.7	13m.9	7.2	458 d	Mira variable (M)
AE Aur	Auriga	5m.78	6m.08	0.3		Orion variable (INA)
R Boo	Boötes	6m.2	13m.1	6.9	223 d	Mira variable (M)
W Boo	Boötes	4m.73	5m.4	0.67	450 d	Semiregular (SRB:)
X Cam	Camelopardalis	7m.4	14m.2	6.8	144 d	Mira variable (M)
VZ Cam	Camelopardalis	4m.80	4m.96	0.16	23.7 d	Semiregular (SR)
R Cap	Capricornus	9m.4	14m.9	5.5	345 d	Mira variable (M)

η Car	Carina	-0m.8	7m.9	8.6		S Doradus (SDOR)
l Car	Carina	3m.28	4m.18	0.9	35 d	Classical Cepheid (DCEP)
R Car	Carina	3m.9	10m.5	6.6	309 d	Mira variable (M)
S Car	Carina	4m.5	9m.9	5.4	149 d	Mira variable (M)
γ Cas	Cassiopeia	1m.6	3m.0	1.4		Gamma Cassiopeiae (GCAS)
R Cas	Cassiopeia	4m.7	13m.5	8.9	430 d	Mira variable (M)
S Cas	Cassiopeia	7m.9	16m.1	8.2	612 d	Mira variable (M)
W Cas	Cassiopeia	7m.8	12m.5	4.7	406 d	Mira variable (M)
WZ Cas	Cassiopeia	6m.3	8m.5	2.2		Semiregular (SRB)
R Cen	Centaurus	5m.3	11m.8	6.5	546 d	Mira variable (M)
S Cen	Centaurus	9m.2 (p)	10m.7 (p)	1.5	65 d	Semiregular (SR)
T Cen	Centaurus	5m.5	9m.0	3.5	90.44 d	Semiregular (SRA)
V645 Cen (Proxima Centauri)	Centaurus	12m.1 (B)	13m.12 (B)	1.02		UV Ceti (UV)
δ Cep	Cepheus	3m.48	4m.37	0.89	5.36634 d	Classical Cepheid (DCEP) prototype
μ Cep	Cepheus	3m.43	5m.1	1.67	730 d	Semiregular (SRC)
S Cep	Cepheus	7m.4	12m.9	5.5	487 d	Mira variable (M)
T Cep	Cepheus	5m.2	11m.3	6.1	388 d	Mira variable (M)
U Cep	Cepheus	6m.75	9m.24	2.49	2.49 d	Eclipsing binary Algol type (EA/SD)
SS Cep	Cepheus	8m.0 (p)	9m.1 (p)	1.1	90 d	Semiregular (SRB)

(continued)

Table 7.1 (continued)

Long Period Variable List

Name	Constellation	Max magnitude	Min Magnitude	Range	Period	Type
AR Cep	Cepheus	7m.0	7m.9	0.9		Semiregular (SRB)
o Cet (Mira)	Cetus	2m.0	10m.1	8.1	332 d	Mira variable (M)
T Cet	Cetus	5m.0	6m.9	1.9	159 d	Semiregular (SRC)
U Cet	Cetus	6m.8	13m.4	6.6	235 d	Mira variable (M)
W Cet	Cetus	7m.1	14m.8	7.7	351 d	Mira variable (M)
R Cha	Chamaeleon	7m.5	14m.2	6.7	335 d	Mira variable (M)
R CMa	Canis major	5m.70	6m.34	0.64	1.13 d	Eclipsing binary Algol type (EA/SD)
VY CMa	Canis major	6m.5	9m.6	3.1		unique (*)
FW CMa	Canis major	5m.00	5m.50	0.5		Gamma Cassiopeiae (GCAS)
S CMi	Canis minor	6m.6	13m.2	6.6	333 d	Mira variable (M)
R Cnc	Cancer	6m.07	11m.8	5.73	362 d	Mira variable (M)
S Cnc	Cancer	8m.29	10m.25	1.96	9.48 d	Eclipsing binary Algol type (EA/DS)
T Cnc	Cancer	7m.6	10m.5	2.9	482 d	Semiregular (SRB)
X Cnc	Cancer	5m.6	7m.5	1.9	195 d	Semiregular (SRB)
T Col	Columba	6m.6	12m.7	6.1	226 d	Mira variable (M)
R Com	Coma Berenices	7m.1	14m.6	7.5	363 d	Mira variable (M)
a Cor Borealis	Corona Borealis	2m.21 (B)	2m.32 (B)	0.11	17.35 d	Eclipsing binary Algol type (EA/DM)

R CrB	Corona Borealis	5m.71	14m.8	9.09		R Coronae Borealis (RCB)
S CrB	Corona Borealis	5m.8	14m.1	8.7	360 d	Mira variable (M)
T CrB	Corona Borealis	2m.0	10m.8	8.8	(80 years)	recurrent nova (NR)
U CrB	Corona Borealis	7m.66	8m.79	1.13	3.45 d	Eclipsing binary Algol type (EA/SD)
V CrB	Corona Borealis	6m.9	12m.6	5.3	358 d	Mira variable (M)
W CrB	Corona Borealis	7m.8	14m.3	6.5	238 d	Mira variable (M)
R Cru	Crux	6m.40	7m.23	0.83	5.82 d	Classical Cepheid (DCEP)
R Crv	Corvus	6m.7	14m.4	7.7	317 d	Mira variable (M)
χ Cyg	Cygnus	3m.3	14m.2	10.9	408 d	Mira variable (M)
R Cyg	Cygnus	6m.1	14m.4	8.3	426 d	Mira variable (M)
U Cyg	Cygnus	5m.9	12m.1	6.2	463 d	Mira variable (M)
W Cyg	Cygnus	6m.80 (B)	8m.9 (B)	2.1	131 d	Semiregular (SRB)
X Cyg	Cygnus	5m.85	6m.91	1.06	16.38 d	Classical Cepheid (DCEP)
RT Cyg	Cygnus	6m.0	13m.1	7.1	190 d	Mira variable (M)
SS Cyg	Cygnus	7m.7	12m.4	4.7	49.5 d	dwarf nova UGSS prototype

(continued)

Table 7.1 (continued)

Long Period Variable List

Name	Constellation	Max magnitude	Min Magnitude	Range	Period	Type
SU Cyg	Cygnus	6m.44	7m.22	0.78	3.84 d	Classical Cepheid (DCEP)
CH Cyg	Cygnus	5m.60	8m.49	2.89		Z Andromedae (ZAND+SR)
R Del	Delphinus	7m.6	13m.8	6.2	285 d	Mira variable (M)
U Del	Delphinus	7m.6 (p)	8m.9 (p)	1.3	110 d	Semiregular (SRB)
EU Del	Delphinus	5m.79	6m.9	1.11	59.7 d	Semiregular (SRB)
β Dor	Dorado	3m.46	4m.08	0.62	9.84 d	Classical Cepheid (DCEP)
S Dor	Dorado	8m.6 (B)	11m.5 (B)	2.9		S Doradus (SDOR) (prototype)
R Dra	Draco	6m.7	13m.2	6.5	246 d	Mira variable (M)
T Eri	Eridanus	7m.2	13m.2	6	252 d	Mira variable (M)
R For	Fornax	7m.5	13m.0	5.5	389 d	Mira variable (M)
η Gem	Gemini	3m.15	3m.9	0.75	233 d	Semiregular (SRA+EA)
ζ Gem	Gemini	3m.62	4m.18	0.56	10.15 d	Classical Cepheid (DCEP)
R Gem	Gemini	6m.0	14m.0	8	370 d	Mira variable (M)
S Gem	Gemini	8m.0	14m.7	6.7	293 d	Mira variable (M)
T Gem	Gemini	8m.0	15m.0	7	288 d	Mira variable (M)
U Gem	Gemini	8m.2	14m.9	6.7	105.2 d	dwarf nova (UGSS+E)
S Gru	Grus	6m.0	15m.0	9	402 d	Mira variable (M)

α Her	Hercules	2m.74	4m.0	1.26		Semiregular (SRC)
g Her (30 Her)	Hercules	4m.3	6m.3	2	89.2 d	Semiregular (SRB)
u Her (68 Her)	Hercules	4m.69	5m.37	0.68	2.05 d	Eclipsing binary Algol type (EA/SD)
S Her	Hercules	6m.4	13m.8	7.4	307 d	Mira variable (M)
U Her	Hercules	6m.4	13m.4	7	406 d	Mira variable (M)
X Her	Hercules	7m.5 (p)	8m.6 (p)	1.1	95.0 d	Semiregular (SRB)
R Hor	Horologium	4m.7	14m.3	9.6	408 d	Mira variable (M)
U Hor	Horologium	7m.8 (p)	15m.1 (p)	6.3	348 d	Mira variable (M)
R Hya	Hydra	3m.5	10m.9	7.4	389 d	Mira variable (M)
S Hya	Hydra	7m.2	13m.3	6.1	257 d	Mira variable (M)
U Hya	Hydra	7m.0 (B)	9m.4 (B)	2.4	450 d	Semiregular (SRB)
VW Hya	Hydra	10m.5	14m.1	3.6	2.69 d	Eclipsing binary Algol type (EA/SD)
BL Lac discovered to be a blazar	Lacerta	12m.4 (B)	17m.2 (B)	4.8		BLLAC prototype
R Leo	Leo	4m.4	11m.3	6.9	310 d	Mira variable (M)
R Lep	Lepus	5m.5	11m.7	6.2	427 d	Mira variable (M)
RX Lep	Lepus	5m.0	7m.4	2.4	~60 d	Semiregular (SRB)
R LMi	Leo Minor	6m.3	13m.2	6.9	372 d	Mira variable (M)
RU Lup	Lupus	9m.6 (p)	13m.4 (p)	3.8		Orion variable (INT)
β Lyr	Lyra	3m.25	4m.36	1.11	12.91 d	Eclipsing binary Beta Lyrae type (prototype)

(continued)

Table 7.1 (continued)

Long Period Variable List

Name	Constellation	Max magnitude	Min Magnitude	Range	Period	Type
R Lyr	Lyra	3m.88	5m.0	1.12	~46 d	Semiregular (SRB)
RR Lyr	Lyra	7m.06	8m.12	1.06	0.566 d	RR Lyrae variable (prototype)
U Mic	Microscopium	7m.0	14m.4	7.7	334 d	Mira variable (M)
U Mon	Monoceros	6m.1 (p)	8m.8 (p)	2.7	91.3 d	RV Tauri variable (RVB)
V Mon	Monoceros	6m.0	13m.9	7.9	341 d	Mira variable (M)
R Nor	Norma	6m.5 (p)	13m.9 (p)	7.4	508 d	Mira variable (M)
T Nor	Norma	6m.2	13m.6	7.4	241 d	Mira variable (M)
R Oct	Octans	6m.4	13m.2	6.8	405 d	Mira variable (M)
S Oct	Octans	7m.2	14m.0	6.8	259 d	Mira variable (M)
V Oph	Ophiuchus	7m.3	11m.6	4.3	297 d	Mira variable (M)
X Oph	Ophiuchus	5m.9	9m.2	3.3	329 d	Mira variable (M)
RS Oph	Ophiuchus	4m.3	12m.5	8.2		recurrent nova (NR)
BF Oph	Ophiuchus	6m.93	7m.71	0.78	4.06 d	Classical Cepheid (DCEP)
α Ori (Betelgeuse)	Orion	0m.0	1m.3	1.3	6.39 years	Semiregular (SRC)
δ Ori (Mintaka)	Orion	2m.14	2m.26	0.12	5.73 d	Eclipsing binary Algol type (EA/DM)
R Ori	Orion	9m.05	13m.4	4.35	377 d	Mira variable (M)
U Ori	Orion	4m.8	13m.0	8.5	368 d	Mira variable (M)
W Ori	Orion	8m.2 (p)	12m.4 (p)	4.2	212 d	Semiregular (SRB)

VV Ori	Orion	5m.31	5m.66	0.35	1.48 d	Eclipsing binary Algol type (EA/KE:)
CK Ori	Orion	5m.9	7m.1	1.2	120 d	Semiregular (SR:)
κ Pav	Pavo	3m.91	4m.78	0.87	9.09 d	Type II Cepheid (CW)
S Pav	Pavo	6m.6	10m.4	3.8	381 d	Semiregular (SRA)
β Peg	Pegasus	2m.31	2m.74	0.43		LB
ε Peg (Enif)	Pegasus	0m.7	3m.5	2.8		LC
R Peg	Pegasus	6m.9	13m.8	6.9	378 d	Mira variable (M)
X Peg	Pegasus	8m.8	14m.4	5.8	201 d	Mira variable (M)
β Per (Algol)	Perseus	2m.12	3m.39	1.27	2.80 d	Eclipsing binary Algol type (EA/SD) (prototype)
DY Persei	Perseus					R Coronae Borealis / DY Persei / (prototype)
φ Per	Perseus	3m.96	4m.11	0.15	19.5 d	Gamma Cassiopeiae (GCAS)
ρ Per	Perseus	3m.30	4m.0	0.7	50 d	Semiregular (SRB)
X Per	Perseus	6m.03	7m.0	0.97		Gamma Cassiopeiae (GCAS+XP)
ζ Phe	Phoenix	3m.91	4m.42	0.51	1.66 d	Eclipsing binary Algol type (EA/DM)
R Pic	Pictor	6m.35	10m.1	3.75	171 d	Semiregular (SR)
R Psc	Pisces	7m.0	14m.8	7.8	345 d	Mira variable (M)
TX Psc	Pisces	4m.79	5m.20	0.42		LB

(continued)

Table 7.1 (continued)

Long Period Variable List

Name	Constellation	Max magnitude	Min Magnitude	Range	Period	Type
L2 Pup	Puppis	2m.6	6m.2	3.6	141 d	Semiregular (SRB)
RS Pup	Puppis	6m.52	7m.67	1.15	41.31 d	Classical Cepheid (DCEP)
T Pyx	Pyxis	7m.0 (B)	15m.77 (B)	8.77	20 years	recurrent nova (NR)
S Scl	Sculptor	5m.5	13m.6	8.1	363 d	Mira variable (M)
RR Sco	Scorpius	5m.0	12m.4	7.4	281 d	Mira variable (M)
RS Sco	Scorpius	6m.2	13m.0	6.8	320 d	Mira variable (M)
RT Sco	Scorpius	7m.0	15m.2	8.2	449 d	Mira variable (M)
R Sct	Scutum	4m.2	8m.6	4.4	146.5 d	RV Tauri variable (RVA)
R Ser	Serpens	5m.16	14m.4	9.24	356 d	Mira variable (M)
S Ser	Serpens	7m.0	14m.1	7.1	372 d	Mira variable (M)
U Sge	Sagitta	6m.45	9m.28	2.83	3.38 d	Eclipsing binary Algol type (EA/SD)
WZ Sge	Sagitta	7m.0 (B)	15m.53 (B)	8.53	33 years	dwarf nova (UGSU+E+ZZ)
RR Sgr	Sagittarius	5m.4	14m.0	8.6	336 d	Mira variable (M)
R Sgr	Sagittarius	6m.7	12m.83	6.13	270 d	Mira variable (M)
U Sgr	Sagittarius (in M25)	6m.28	7m.15	0.87	6.74 d	Classical Cepheid (DCEP)
RT Sgr	Sagittarius	6m.0	14m.1	8.1	306 d	Mira variable (M)
RU Sgr	Sagittarius	6m.0	13m.8	7.8	240 d	Mira variable (M)

RY Sgr	Sagittarius	5m.8	14m.0	8.2		R Coronae Borealis (RCB)
VX Sgr	Sagittarius	6m.52	14m.0	7.08	732 d	Semiregular (SRC)
λ Tau	Taurus	3m.37	3m.91	0.54	3.95 d	Eclipsing binary Algol type (EA/DM)
R Tau	Taurus	7m.6	15m.8	8.2	321 d	Mira variable (M)
T Tau	Taurus	9m.3	13m.5	4.2		Orion variable (INT)
SU Tau	Taurus	9m.1	16m.86	7.76		R Coronae Borealis (RCB)
R Tri	Triangulum	5m.4	12m.6	7.2	267 d	Mira variable (M)
R UMa	Ursa Major	6m.5	13m.7	7.2	302 d	Mira variable (M)
T UMa	Ursa Major	6m.6	13m.5	6.9	257 d	Mira variable (M)
U UMa	Ursa Major	6m.20	6m.25	0.05		-
W UMa	Ursa Major	7m.75	8m.48	0.73	0.33 d	Eclipsing binary W Ursae Majoris type (prototype)
Z UMa	Ursa Major	6m.2	9m.4	3.2	196 d	Semiregular (SRB)
R Vir	Virgo	6m.1	12m.1	6	146 d	Mira variable (M)
S Vir	Virgo	6m.3	13m.2	6.9	375 d	Mira variable (M)

Chapter 8

Cepheids and Other Variable Types

Variable stars may have been known since antiquity, but one particular type of star is all important in the multiplicity of applications to our understanding of the universe. How and why some stars vary is a study that was essentially started by the Cepheid variable stars, a group of objects bunched under the name of their progenitor, δ in the northern constellation of Cepheus the King.

We have seen in earlier chapters the history and development of our understanding of these stars from John Goodricke and Edward Piggott in the 18th century to Ejnar Hertzsprung and Arthur Eddington in the 20th. The cyclic variability and absolute assurance of regularity exhibited by these stars has become a byword in astronomy, and the application to discovering the distance scale of the universe by means of Cepheid variables is a tale in itself that was partly covered in Chapter Two. Edwin Hubble used Cepheid variables to both discover and to calibrate his distances to galaxies in the 1920's, and the evidence he uncovered enabled George LeMaitre to postulate that the entire universe came from a primordial atom, an idea we now term the Big Bang. They are a very versatile group of variable stars!

Watching these regular variable stars is how many observers cut their teeth. The reliability of such stars enables astronomers to establish their skills in drawing light curves, in estimating their magnitude differences and in recognizing the fields of view and comparisons that need to be made as part of amateur scrutiny. From a professional point of view, the Cepheid instability strip on the HR diagram underscores the processes that occur in

© Springer Nature Switzerland AG 2018

M. Griffiths, *Observer's Guide to Variable Stars*, The Patrick Moore
Practical Astronomy Series, https://doi.org/10.1007/978-3-030-00904-5_8

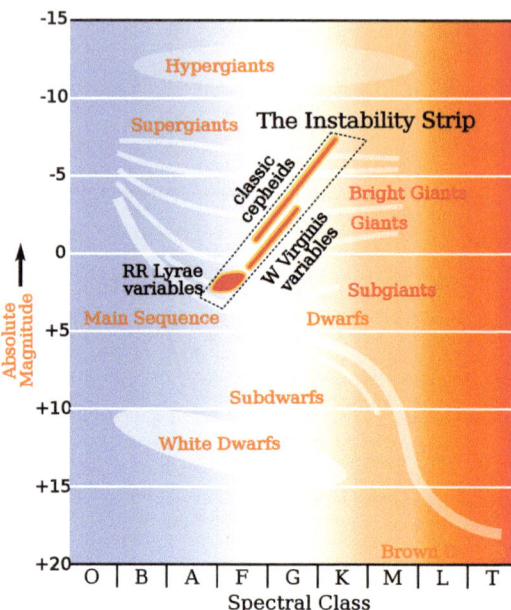

Fig. 8.1 Cepheid instability strip (Image from https://sites.google.com/site/aavso-sppsection/spp-star-classes.)

stars of a certain age and mass and enable astronomers to predict behavior beyond the main sequence very well. The Cepheid strip can be seen here in Fig. 8.1 as a reminder of its location and importance. Note the confined location of the strip in the form of the star's luminosity and spectral types.

The RR Lyrae, W Virginis, δ Cephei, δ Scuti and ZZ Ceti stars all lie in this narrow portion of the diagram, denoting a specific range of reactions and stellar functions as the stars attempt to regain their foothold on the main sequence after the onset of helium fusion in the core.

Cepheid variable stars are radially pulsating, high luminosity giant stars with MKK classes of Ib to II, with periods that lie within the broad range of 1 to 135 days and have amplitudes from a few hundredths of a magnitude to over 2 magnitudes visually. As they pulsate they change their spectral types from F at maximum light to G at minimum light output and occasionally dipping to K, depending on type. As a rule of thumb, the longer the period of variability, the later is the spectral type. If we could see the star in its entirely we would see a huge sphere pulsing in and out, almost breathing, and due to the luminosity function, of course, the maximum light output corresponds to the largest diameter and surface area.

Cepheid variables have a few subtypes that are worthy of note, including some that are very difficult to discriminate due to light output, spectral type and period range. However, spectroscopy gives us the tools to separate them out into distinct groups and evolutionary stages. For example, a typical spectral difference between W Virginis stars and classical Cepheids is the presence of hydrogen-line emission in the former and of Ca II H and K emission in the latter. Therefore, we can understand that the outer atmospheres of these stars are revealing underlying physical differences between giant and subgiant stars.

Observing such stars and their regularity is a bonus for observers, as many are bright, easily recognized and there have been many stars already studied with which to make comparisons. As stated earlier, observing Cepheid variables is a very good introduction to the world of variable stars.

The following section will list the main variables and their differences.

ACEP

The term ACEP stands for Anomalous Cepheids. These are stars with periods characteristic of comparatively longer periods of variability that range from a few hours to a few days but should not be confused with RR Lyrae types, as these stars are brighter and more massive, generally falling within the range of 1.3 to 2.2 solar masses in comparison to the RR Lyrae mass of 0.5 solar masses on average. Some of these ACEOP types are relatively metal poor, indicating that they have evolved in the distant past and have smaller metallicities than sun-like stars.

CW

These are the classical W Virginis-type variable. W Virginis stars are bodies that, as Population II stars, are part of the spherical halo of our galaxy and therefore old, metal-poor stars with small amplitudes and shorter periods. In the case of many CW stars, they have periods of variability lasting between 0.8 and, in extreme cases, up to 35 days, and most have amplitudes in light output that cover the range from 0.3 to 1.2 magnitudes visually.

W Virginis stars have a period-luminosity relation slightly different from that for δ Cep variables. In comparison classical Cepheids and W Virginis types reveal that their periods are matched in time, and the CW stars are fainter by an amount that varies from 0.7 to 2 magnitudes. Instead of having a smooth decline, too, some CW stars have a flat line profile in the light curve as they decline to minimum, indicating that the pulsation size between maximum and minimum is relatively even. CW stars can be found in globular clusters as well as the galactic halo, where comparisons can be made between them and RR Lyrae types.

It is important to recognize that W Virginis types can be further defined as A and B types where they have the categorization CWA and CWB. The differences are due to the length of the variable period. In CWA stars the variable period is longer than 8 days, whereas CWB stars have periods shorter than 8 days. Some short-period CWB stars are also known as BL Herculis variables, after the progenitor star.

DCEP

The DCEP stars are *the* classical Cepheid stars, or δ Cephei-type variables. They are comparatively young galactic objects that have left the main sequence and evolved into the instability strip of the H-R diagram, and they pulsate due to the κ mechanism discovered by Arthur Eddington. These stars follow the typical Cepheid period-luminosity relationship discovered by Henrietta Leavitt and can be used as excellent distance indicators out to a approximately 100 million light years if they can be imaged. In 2001, the distance scale and the Hubble law were calibrated using Cepheid variable stars during the Hubble Key Project. The study showed that the Hubble constant had a value of 72 km/s/Mpc, a value since refined by other studies such as WMAP to 68 km/s/mpc; so the value of Cepheids as distance indicators is very stable and strong evidence of confirmation of our knowledge of galactic distances.

Classical Cepheids belong to the Population I stars of the galactic disk and can be found even in some open clusters. Their light curves are very distinctive, with a rapid rise followed by a longer tail as can be seen from Fig. 8.2 here.

The classical Cepheids also have a small range of subtypes that are dependent on the pulsation mode. The DCEP(B) stars have two modes operating at the same time and vary over periods of between 2 and 7 days but not longer.

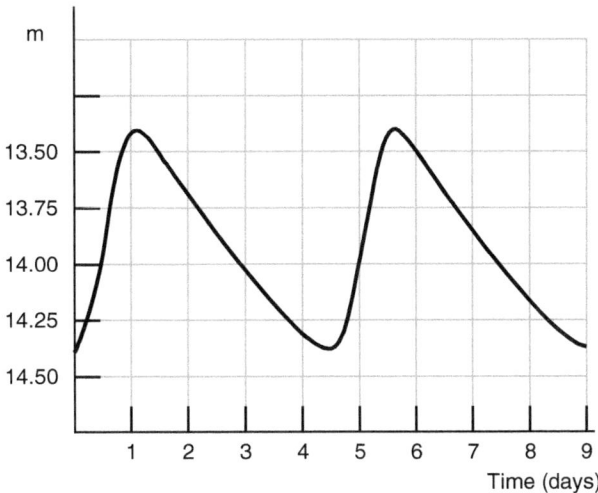

Fig. 8.2 Cepheid light curve (Image from https://www.quora.com/What-are-cepheid-variables.)

DSCT

These are variables of the well-known d Scuti type, which are pulsating variables of spectral types A0-F5 and range from subgiants of Type III to main sequence stars of V type as the instability strip descends to touch the main sequence at its lowest point among the A- and F-type stars. Delta Scuti stars have smaller light amplitudes than the classical Cepheids, with differences varying between a few thousandths of a magnitude up to a magnitude.

Studies of these stars and their light curves quickly reveal how distinct each one is, as hardly any of the light curves are representative of the type! They reveal that both radial and non-radial movement is present on the star, and in more than one case, a δ Scuti star has stopped pulsating altogether. DSCT stars appear to be Population I-type objects of the galactic disk, and as expected, there are several subgroups among them: the type DSCT(C) having the lowest magnitude amplitude among the groups, while the HADSCT type stars have higher amplitudes. δ Scuti stars are often found in open clusters scattered across the Milky Way.

GDOR

These variables are named after their progenitor type γ Doradus in the southern hemisphere sky. The main sequence dwarfs and subgiants with MKK classes of Iv and V with spectral types typically of late A to late F. They have very small amplitudes which have not been observed to exceed 0.1 magnitudes, and the stars have a typical period range from 0.3 to 3 days. Their low magnitude amplitude makes them difficult objects of study for the amateur unless photometry is undertaken.

L

It is difficult to be precise about where the L-type variables lie in any study as they reveal complex light curves and very little evidence of regular periods. They are sometimes titled slow irregular variables, but this epithet hardly encourages observation! Within the world of variable star observation, the L-type stars are lumped together as being simply insufficiently studied, and they could well belong to known variable types if more long term studies of them were available. This obviously encourages an amateur study of such stars, and the AAVSO will have some indicators of where to find such stars. We shall meet them in the chapter on long-period variables.

Given that many Cepheid variable stars have significant brightness, they can be avidly studied by the inexperienced who wish to go on after some practice to greater and more challenging variable star work. These regular beacons in the sky are a fantastic group of stars in that the classical Cepheids have the regularity needed for the beginner and other classes still retain the challenge for more detailed and vital observational work in order to advance our knowledge of these types. If one wishes to take up this kind of study, then local societies may have access to experienced observers to encourage you, and international bodies such as the AAVSO and the BAA will have the charts and comparisons to enable you to make a great start in this vital field.

For those observers who wish to follow these unusual stars, the BAA variable star section also has an exhaustive list of variables that can be followed by amateur observers. These important fields of work are detailed at the following URL: http://www.britastro.org/vss/.

Chapter 9

Rotating Variable Stars

In 1612 Galileo Galilei discovered that the Sun rotates on its axis. He worked this out by watching the movement of sunspots across the solar disk and calculated the solar rotation period to be approximately 25 days on average. For the first time, a body such as a star was found to be imperfect and blemished, and this led over the next two centuries to speculation regarding the nature of some variable stars. Did they vary due to star spots? Was their rotation similar to that of the Sun? Indeed, do other stars have spots, and if so, are the mechanisms that cause them similar to those of our own Sun?

Initial observations of variable stars turned on such questions, and many early observers, such as Fabricius, Piggott and Goodricke, initially believed that variability was due to spots and stellar rotation. These ideas were rapidly dispelled by observation and the application of growing sciences. Nevertheless, there remain some rather interesting variable stars that change due to star spots, or due to rotation, as they are not completely spherical. The AAVSO recognizes several types, and we shall discuss each type below. It must be remembered that the changes in luminosity that we observe with such stars are not due to physical changes but in most cases are simply due to their rotation.

Rotating ellipsoidal variables, known by the appellation ELL, are close binary systems whose components are ellipsoid in shape. They are not eclipsing in the usual sense, as it seems unlikely that the components are in

© Springer Nature Switzerland AG 2018
M. Griffiths, *Observer's Guide to Variable Stars*, The Patrick Moore Practical Astronomy Series, https://doi.org/10.1007/978-3-030-00904-5_9

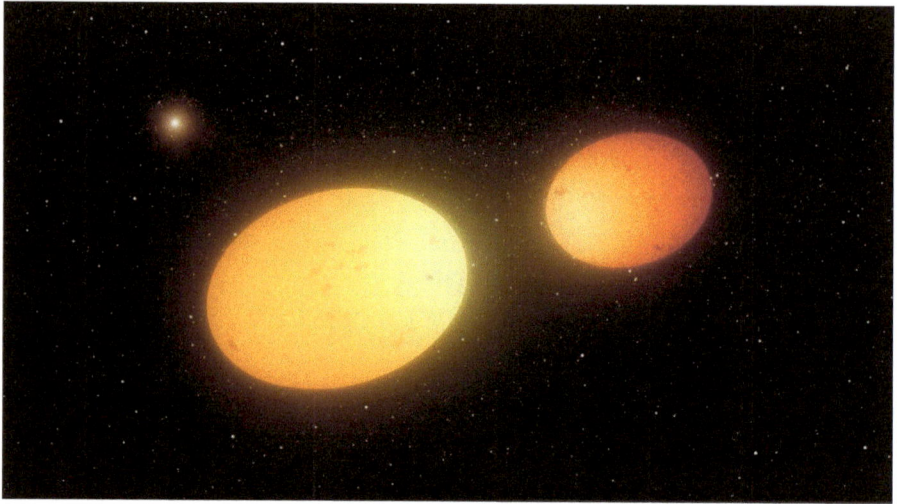

Fig. 9.1 Ellipsoid variable. (Image from https://www.thoughtco.com/heartbeat-stars-overview-4105965.)

our plane of sight, but their fluctuations occur due to changes in the amount of light-emitting area that is visible to the observer.

Rotating ellipsoidal variable stars demonstrate some of the strange qualities of our stellar neighbors. We are familiar with the idea of stars with spherical shapes, but it comes as a shock, and a stretch of the imagination, when imaging stars, to see ones that are egg shaped! Fig. 9.1 shows these stars as imagined by an artist.

These properties demonstrate that stars have enormous gravities that bind the gas, but the gaseous construction allows them some fluidity, especially in the close confines of some binary star systems so that they lose their shape somewhat but still retain a gravitational hold, for the most part, over their gaseous structure.

In such variables, the luminosity fluctuation is usually quite small; most do not exceed 0.1 of a magnitude, so observing such variables generally requires either an expert eye or standard photometric methods. Although most ellipsoidal and rotational variables do not have an excessive range, they are nevertheless important as they demonstrate the physical charac-teristics of some systems. An example of a typical rotating ellipsoidal vari-able is the star Spica, the brightest star in the constellation Virgo. The components orbit each other every four days and are so close that both the primary and secondary star are pulled out of a spherical aspect into egg shapes that are almost touching each other. Obviously their differences in light output vary over this four-day period.

The common AAVSO classes of such stars as used internationally are displayed and described herein in order of alphabetical type.

Rotating Ellipsoidal Variables

α Canum Venaticorum (ACV types)

Actually the star under scrutiny is the companion of Cor Caroli: α2 Canes Venatici. These variable stars are main sequence stars with spectral types B8p to A7p, which also display the characteristic of very strong magnetic fields. Astrophysical studies have shown that the ACV stars typically exhibit magnetic field and brightness changes together, although the amplitudes of the brightness changes are very small, falling within 0.01 to 0.1 magnitude in visible light. Such stars have a large range of periods, too, from 0.5 days to over 160 days within the class.

BY

BY Draconis-type variables are emission-line emitting red dwarfs of dKe to dMe spectral type showing irregular but almost periodic light changes with periods from a fraction of a day to 120 days and amplitudes from several hundredths of a magnitude up to 0.5 magnitude in the visible range. The light variability is caused by axial rotation of a star with a variable degree of non-uniformity of the surface brightness (possible spots?) and chromospheric activity. Some of these stars also show flares similar to those of UV Ceti stars, and in those cases they also belong to the latter type and so are also considered to be eruptive variables.

CTTS/ROT

Classical T Tauri stars showing periodic variability due to spots. Obviously the birth of stars is attended by various irregularities in light output, and many of these types show aspects of variability that fall within several classes. Apparently, these classical T Tauri stars have extensive equatorial disks that result in strong emission lines and emulate the INT and IT variable types.

ELL

Rotating ellipsoidal variables are classical close binary systems with ellipsoidal components that change their combined brightness with periods being equal to those of their orbital motion. The stars then vary for the simple reason that their luminous area changes as they rotate toward or away from the observer, but for the most part such stars offer no eclipse patterns in the light curve. Light amplitudes usually do not exceed 0.1 mag. in visible light.

FKCOM

FK Comae Berenices-type variables are rapidly rotating single G- and K-type giants with non-uniform surface brightness. Their photometric behavior is similar to that of RS CVn systems, but studies have shown that their absorption lines exhibit rotational broadening, with large velocities of 100 to 160 km/s. They have strong magnetic field activity, and some of the class exhibit X-ray emission. FKCOM stars apparently contain some of the hottest coronal plasmas among active stars. Periods of light variation (up to several days) are again equal to the rotational periods, and their light curves reveal amplitudes of several tenths of a magnitude. Some astronomers consider this class to be the product of the further evolution of W Uma-type variables and may be stars that are beginning to coalesce.

LERI

λ Eri type variables are Be stars with the light variation caused by rotational modulation or non-radial pulsations. Their light curves are marked with changing amplitudes, but their magnitude range is not large. Such stars usually have variable periods of the order of 0.3 to 3 days in total.

PSR

These are a very challenging object for amateurs and are a rare phenomenon with inherent difficulties of study due to their faintness. PSR type variables are in fact variable pulsars, which are visible in natural light and of course the light is generated by rapidly rotating neutron stars with extremely strong magnetic fields. Pulsars, first discovered by Jocelyn Bell-Burnell in 1963, are known to emit narrow beams of radiation, and their periods and light curves therefore coincide with this rotation. As pulsars are very rapid rotators, some of the periods are no more than a few thousandths of a second! This would be a very hard object to observe, especially as the amplitudes are generally small, too, somewhere between 0.5 and 0.1 of a magnitude.

R

R-type ellipsoidal variables are generally thought to be close binary systems characterized by the presence of strong reflection of the light of the primary star, usually a hot star that then illuminates the surface of a cooler companion. Light curves of these R types are sinusoidal, with the period equal to the orbital period and maximum brightness coinciding with the passage of the hotter star in front of the companion. In such stars, eclipses are not always absent, but there is no guarantee that all in the class will eclipse. R-type stars show a range in magnitude – up to a whole magnitude in some cases in visible light.

RS

A significant property of RS Canum Venaticorum-type binary systems is the presence in their spectra of strong lines of variable intensity that occur in the chromosphere. These stars, unlike the Sun at such distances, are also characterized by strong radio and X-ray emission. Their light curves look like sine waves, with amplitudes and positions that change with time. The reason for this slow change in time has been explained by means of differential rotation of the star. If its surface is covered with groups of spots, then the period of the rotation of a spot group is usually close to the period of orbital motion. However, this differential rotation changes slowly over time, and the light curve shows the slow migration of the phases of the minimum and maximum light output. Fig. 9.2 demonstrates the possible appearance of these stellar spots.

The variability of star may be up to 0.5 of a magnitude in some cases, and this variability is generally thought to be due to a long-period stellar activity cycle similar to the 11-year solar activity cycle. Just as with our Sun, the surface activity of sunspots, flares and prominences does not remain constant but is continually varying. RS stars are a fascinating type

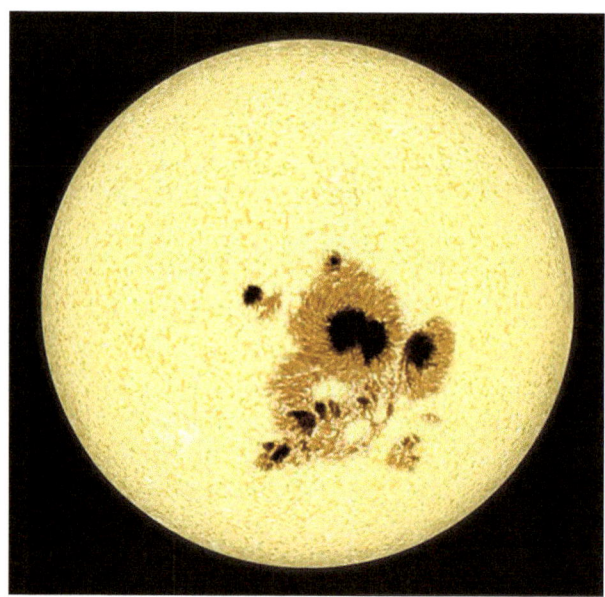

Fig. 9.2 Huge stellar spots (Image from http://www.bbc.co.uk/science/space/ solarsystem/solar_system_highlights/solar_cycle.)

of variable and are important, as the similarity in physical characteristics to our Sun reveal that stars all have similar processes to our parent star and that these processes can be studied and understood.

SXARI

SX Arietis-type variables are main sequence B0p-B9p stars with variable-intensity absorption lines of helium and silicon and are characterized by extreme magnetic fields. SXARI types are sometimes called helium variables. As with all ellipsoidal variables, the periods of light changes coincide with their rotational periods. Their amplitudes are challenging, as generally such stars do not vary much beyond 0.1 of a magnitude, thus making photometry the obvious choice to study them.

WTTS/ROT

These faint but interesting stars are characterized by weak spectral lines and are probably T Tauri-type stars that vary due to star spot production. It is thought that these variables lack strong stellar winds and that their

Fig. 9.3 Naked T Tauri (Image from https://www.pinterest.co.uk/pin/176766354104742565/.)

circumstellar accretion disk is also absent, perhaps an indicator that such stars are now settling toward the zero age main sequence. WTTS/ROT stars have been called "naked" T Tauri types, as they seem to have little evidence of the dust clouds and disks that enclose the classical T Tauri types, though they have been known to have enormous flares from time to time. Fig. 9.3 shows this type of "naked" star.

Obviously, many of the above variable types are going to be hard to distinguish as variable stars even by eagle-eyed observers. Such stars will require photometric study to produce a light curve, and some may be beyond the capability of many observers. Nevertheless, rotating and elliptical variable stars are a potential adjunct to an experienced amateur's observing program.

Chapter 10

Following the Light – Eclipsing Variable Stars

Eclipsing variable stars are quite simply a pair of stars revolving about their common center of mass in an orbit whose equatorial plane passes through or very near Earth. An observer on Earth thus sees one member of the binary pass periodically in front of the face of the other and diminish its light output by means of an eclipse. One of the first eclipsing variable stars studied, and was also noted in antiquity, was the bright star Algol in the constellation of Perseus. The first person to accurately work out why such dips in light output were being observed was John Goodricke, the remarkable deaf-mute astronomer who worked out the process in 1782.

Thousands of eclipsing variable stars are now known. They are extremely important due to the fact that their orbital motions allow astronomers to measure differences in light output, spectra and types that then allow us to determine the mass and orbital parameters of both stars.

As the stars orbit their common center of mass, a primary eclipse will occur when the brighter star is eclipsed by the fainter star. A secondary eclipse will then occur when the fainter star is eclipsed by the brighter star. If the two stars are of the same brightness, then both eclipses will be equal, and a dip in the light output will be seen during eclipse.

If one of the stars is much fainter than the other, then the secondary eclipse may not be readily observed by visual observers. If the orbits of the two stars are circular, then the secondary eclipse will occur midway between primary eclipses. If the orbits are elliptical the secondary eclipse may occur earlier or later than midway.

© Springer Nature Switzerland AG 2018
M. Griffiths, *Observer's Guide to Variable Stars*, The Patrick Moore Practical Astronomy Series, https://doi.org/10.1007/978-3-030-00904-5_10

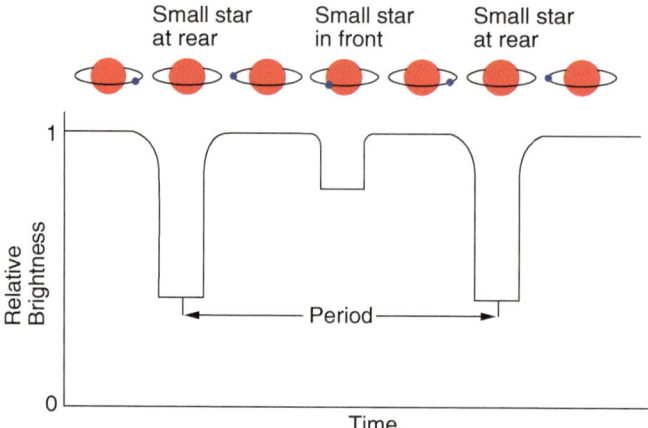

Fig. 10.1 Eclipsing binary light curve (Image from http://webs.mn.catholic.edu.au/physics/emery/hsc_astrophysics_page3.htm.)

A primary eclipse is said to be total if one star is completely obscured by the other star for a period of time. The period of minimum light can be measured in minutes or days. If a star is not totally obscured, then the eclipse is described as partial. In that case there is no prolonged period of minimum light. Figure 10.1 shows a typical light curve of an eclipsing binary star.

EA

EA type systems follow the primary star of their type, β Persei or Algol. Such stars in this class are binaries with spherical or slightly ellipsoidal components. With high precision astronomers can specify from their light curves the moments of the beginning and end of the eclipses in a very precise and accurate fashion. In between eclipses, the light output remains almost constant or, more accurately, varies insignificantly because of the small amount of reflection effects, the slight ellipsoidal shape of the components, or small physical variations. In some eclipsing variables, secondary minima in the light curve may not be apparent at all. However, these types show an extremely wide range of periods, from 0.2 days to in excess of 10,000 days! Light amplitudes are also varied between each type of star and may reach several magnitudes in many cases.

EB

EB variables are the typical beta Lyrae systems. These are eclipsing systems having ellipsoidal components and light curves for which it is impossible to specify the exact times of onset and end of eclipses because of a continuous change of the system's apparent combined brightness between eclipses and the presence of gas swirling around the system. A secondary minimum is observed in all cases with its depth usually being considerably smaller than that of the primary minimum. In the case of EB stars, their variable periods are longer than 0.5 days, and their components are generally spectral types B and A. These stars have light amplitudes during this period which can vary by up to 2 magnitudes in the visual range. Figure 10.2 illustrates the light curve of such close eclipsing systems.

EP

The EP stars are a relatively new group recognized by the IAU, as such systems are actually undergoing eclipses due to the presence of planets. Obviously, the difference in light output is going to be very small and hard

Fig. 10.2 Beta Lyrae light curve (Image from http://www.britastro.org/vss/.)

to record, and even astronomers who are very experienced in photometry may have problems simply differentiating light variability from effects in our own atmosphere. Nevertheless, this is an interesting and important field of recent research.

EW

The EW stars have the progenitor W Ursae Majoris eclipsing. These are eclipsing stars with periods usually shorter than a day, consisting of ellipsoidal components almost in contact and having light curves for which it is generally not possible to specify the exact times of onset and end of the eclipse. Irritatingly, the depths of the primary and secondary minima are almost equal or differ insignificantly. These stars are of the mid-range of spectral types, usually F, G or K, and their light output varies in the visual around 0.5 of a magnitude.

As many of the eclipsing variables are stars that are almost in contact with one another with ellipsoid surfaces, several subgroups have been accepted that illustrate the processes occurring in each system. Some stars are main sequence objects while others are evolving or have evolved away from the main sequence and may or may not be filling the area where materials can be exchanged through gravitational interaction. This area is known as the Roche lobe.

The Roche lobe is the region around a star in a binary system within which orbiting material is gravitationally bound to that star. It is an approximately teardrop-shaped region with the apex of the teardrop pointing towards the other star in the system. When a star evolves, it may grow to such a size that it now exceeds its Roche lobe; its physical surface extends out beyond this gravitational boundary, and the material that lies outside the Roche lobe can be pulled toward a binary companion. As stars evolve in binary systems, mass transfer via Roche lobe overflow occurs and subtly changes the potential of the stellar system. Such transfer is seen in X-ray binaries and also in low mass systems and, in some cases, where the companion is a white dwarf star, leads to recurrent novae.

Detached binary systems usually have EA-type light curves, although some have EB types if both stars are ellipsoidal and close to their Roche lobe limits.

AR

These are stars that remain detached from each other but are evolving off the main sequence as subgiants but have not yet filled their Roche lobes. They are named AR after the progenitor AR Lacertae and show minor variations in magnitude over short periods.

D

Eclipsing variables that remain detached with components not filling their inner Roche lobes.

DM

Differing from the above types in that the components here are main sequence stars that have not filled their inner Roche lobes.

DS

DS-type variables are stars in eclipsing systems where both stars are still detached from each other but one component is a subgiant star that has not yet filled the Roche lobe.

GS

GS variable stars are systems with one or both giant and supergiant components, though one of the components may be a main sequence star.

K

K-type variables are eclipsing binaries that are contact systems with both components filling their inner critical surfaces. They resemble W Ursae Majoris types, as both components may be ellipsoids. They may also be spectral types of K or M.

KE

Similar to the K types with the difference that these variables are O or A spectral type with both components being close in size to their Roche lobes.

WD

Eclipsing stars that are binary systems with at least one white dwarf component, or a single rotating white dwarf. These generally have low visual changes and small variable periods.

W Ser

W Serpentis is the prototype of a class of semi-detached binaries that includes a giant or supergiant transferring material to a massive more compact star. They are characterized by, and distinguished from, the similar β Lyrae systems by strong UV emission from accretion hotspots on a disc of material instead of having swirling gaseous materials surrounding both stars.

Eclipsing binary stars are fascinating to observe and can be appreciated better once the observer has some knowledge of the cosmic dance being played out in such systems. A list of easily observable variable stars is included later in this book.

For those observers who wish to follow these unusual stars, the BAA variable star section has an exhaustive list of variables that can be followed by amateur observers. These important fields of work are detailed at the following URL: http://www.britastro.org/vss/.

Chapter 11

Explosive and Eruptive Variable Stars

The breadth of activity among variable stars is almost overwhelming, and the possibilities of observing them seem almost endless. Many variables are followed by observers for their relatively slow pace of decline and recovery and enable one to gauge smooth light curves and collect data for a number of years. Though patience and dedication are requirements for any variable star work, the observations of eruptive or explosive variable stars is a time consuming and onerous task with no guarantee of results unless time and great effort is given to the task of observing them constantly. Some eruptive variables are fairly consistent, but the majority of them are not, and therefore they require constant scrutiny.

An eruptive variable is a star that undergoes sudden and marked changes in brightness due to activity above the photosphere of the body, usually in the chromosphere or corona. The brightness variations may also be accompanied by an enhanced stellar wind or the ejection of shells of matter. These all add to the mystique of the star and affect the light curves in varied ways.

Early Type Eruptive Variable Stars

There are many different types of eruptive variable star, but the common AAVSO classifications will give the observer some idea of the types of stars we are dealing with. Among the types of eruptive variable are:

© Springer Nature Switzerland AG 2018
M. Griffiths, *Observer's Guide to Variable Stars*, The Patrick Moore
Practical Astronomy Series, https://doi.org/10.1007/978-3-030-00904-5_11

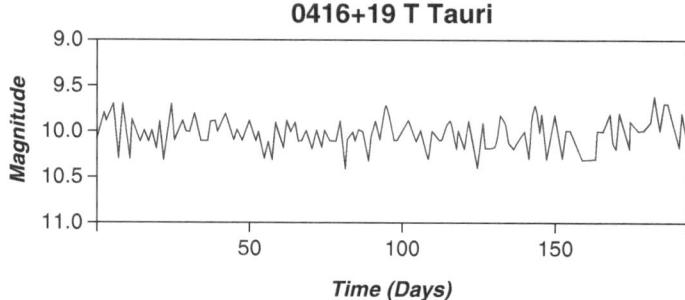

Fig. 11.1 T Tauri light curve (Image from https://www.aavso.org/vsots_ttau.)

Classical T Tauri stars (CTTS)

These have extensive disks and shells of material surrounding them that come from their formative processes. The surrounding material can vary in light output as the star varies. They generally vary rapidly and irregularly over a timescale of days and weeks and generally stay within a 1 or 2 magnitude range as can be seen from Fig. 11.1 here:

FU Orionis

Although T Tauri remains one of the significant stars in its range, the most commonly studied star is FU Orionis. Variables of this type (FU Ori) are characterized by a unique major gradual increase in brightness by about 4 to 6 magnitudes as can be seen from Fig. 11.2. Following this, they show a complex spectrum much like that of an F- or a G-type supergiant star. They then may stay constant at maximum brightness or decline slowly by 1 to 2 mag. several months after the initial rise. One of their relatively unique features is that the great majority of all known FU Ori types are coupled with reflecting comet-like nebulae. Their underlying mechanism is thought to be the transfer of materials from a larger F or even B spectral-type pre-main sequence star to a smaller, close T Tauri-type companion.

EXor

Named after EX Lupi, and collectively known in the literature as EXors, these are eruptive T Tauri stars that show brightening episodes of several magnitudes in time scales of several months or a few years. The EXor stage

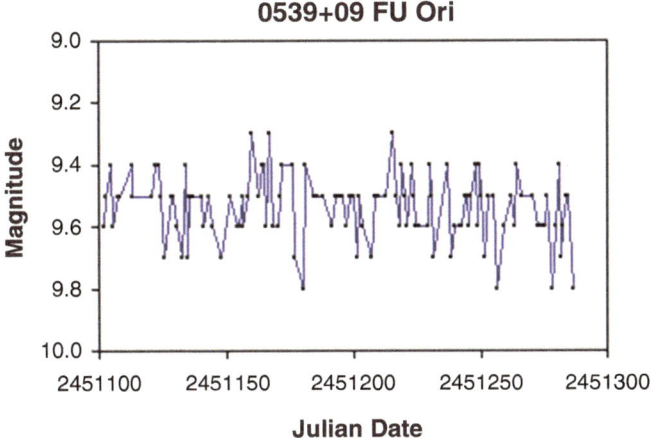

Fig. 11.2 FU Orionis light curve (Image from https://www.aavso.org/vsots_fuori.)

appears to follow the FU Ori one, more than just being a less evident manifestation of the same phase. They are less luminous and present different emission-line spectra than those of FU Orionis, but their light curves show similar patterns.

γ *CAS*

These are eruptive irregular variables thought to be rapidly rotating O or A subgiant stars with mass outflow from their equatorial zones. The formation of the equatorial ring or disk is accompanied by a temporary brightening. (These could fluctuate by up to 1.5 magnitudes.) γ Cassiopeia has remained almost constant since the 1970's, suggesting the star may have settled into some type of equilibrium, but no one is quite sure of this. As can be seen from Fig. 11.3, the light curve shows regularity, with the uneven data sets perhaps indicating loss of data rather than an actual fade.

IN

IN stars are generally known as Orion variables. They can be observed as irregular, eruptive variables connected with bright or dark diffuse nebulae. Some of them show cyclic light variations that may be caused by axial rotation. It is thought that these variables are probably young objects that are

Fig. 11.3 Gamma Cassiopeia 2014 light curve (Image from http://www.britastro. org/vss/.)

evolving toward the zero age main sequence, and their range of brightness variations may reach several magnitudes. They are subdivided into INA and INB types that show regular fading and recoveries. These subtypes are all observed in nebulae across the Milky Way and also fall into further subgroups such as INT, which show rather more rapid variation as the variable process is probably occurring in the chromospheres of INT stars rather than in surrounding materials. For a full description of eruptive variables in nebulae you can look up the types in the AAVSO catalogue. Many of these types are faint and, being irregular, may require an experienced observer to watch them.

UVN

UVN stars are Orion variables of spectral types Ke-Me that show irregular flare activity almost identical to UV Ceti-type variables. In addition to being related to nebulae, they are normally characterized by greater luminosity and slower development of flares. Latest research suggests that they are possibly a specific subgroup of **INB** variables with irregular variations superimposed by flares.

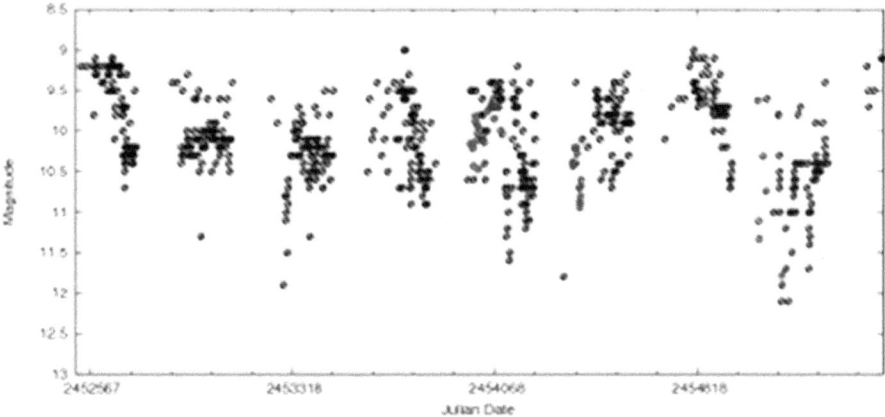

Fig. 11.4 UX Orionis light curve (http://iopscience.iop.org/article/10.1088/0004-6256/149/3/108.)

UX Ori

UX Orionis stars, are a subgroup of young stars that show irregular varia-tions with a wide range of amplitudes from barely detectable to more than 4 magnitudes. Most of them are infrared Herbig Ae/Be-type stars, but there are also some T Tauri stars with later spectral types also showing the same behavior. Variables that show the greatest range in amplitude are giant stars later than B8. There appear to be two principal components to their vari-ability: irregular variations on timescales of days around a mean brightness level and a second component that reveals occasional episodes of deep minima, occurring irregularly but more frequent near the low points of the brightness cycles. Again, these stars are evolving toward the zero-age main sequence. Figure 11.4 gives a typical representative light curve of UX Orionis.

Luminous Blue Variables

In previous years these were commonly called variables of the S Doradus type. Research has shown that these are highly luminous, eruptive, B- to F-type stars that show peculiar structures in their spectrum (and are often subtitled pec for peculiar), and as they are predominantly bright and blue in color, they have been termed luminous blue variables, LBV for short.

These stars show irregular light changes with amplitudes in the range of 1 to 7 mag. And, as their name suggests, they belong to the brightest blue stars of their parent galaxies. Although they may be on the main sequence, these variables are still connected with diffuse nebulae and surrounded by expanding envelopes. The famous star P Cygni is a typical example of an LBV, as is the well-known eruptive southern hemisphere object η Carina. These are stars in which radiation is creating pulsations in the envelope and mass loss and shell ejection is common to all types. They are closely allied to the next set of eruptive variables.

WR

Eruptive Wolf-Rayet variables are stars with the broad emission features of He I and He II, showing that they have hot atmospheres. WR stars often display small, irregular light changes with amplitudes only up to 0.1 of a magnitude, which are probably caused by mass outflow from their atmospheres. Some WR stars also show binary companionship.

Late-Type Eruptive Variable Stars

Just as there are eruptive stars of early types often associated with nebulae, there are late-type eruptives that are a little less common, as such activity is not a feature of stars undergoing the last stages of evolution. In some cases they are almost "reverse"-type variables in that they exhibit fading rather than brightening. The main variable star types of this class can be seen following.

UV Ceti

Eruptive variables of the UV Ceti type are main sequence KVe to MVe stars that occasionally undergo flare activity, with amplitudes from several tenths of a magnitude up to 6 magnitudes. As the process cannot be predicted in advance, observation of UV Ceti-type flare stars has become a bit of a holy grail for amateurs, and monitoring of them is an almost exclusive amateur activity that requires patience and luck! The maximum rise in light output is attained in several seconds after the beginning of a flare, before the star returns to its normal brightness in just a few minutes. This behavior makes

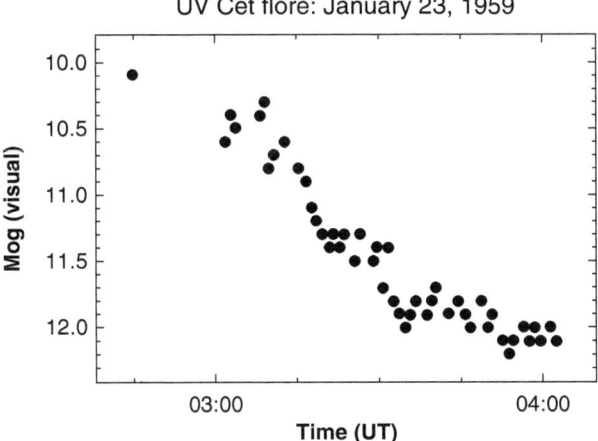

Fig. 11.5 UV Ceti flare light curve (Image from https://www.aavso.org/vsots_ uvceti.)

UV Ceti-type variables especially challenging, as an observer in Japan may catch a flare but an observer in Europe or America will certainly miss the outburst! Figure 11.5 shows a typical light curve.

UV Ceti stars are not strictly late-type eruptives, but as they are red dwarfs that have been on the main sequence for billions of years they can be considered to be old stars. The mechanism behind the outburst is commonly thought to be intense magnetic activity akin to solar flares.

RCB

These stars are variables of the R Coronae Borealis type and are hydrogen-deficient, carbon- and helium-rich luminous stars of spectral types Bpe to C. They are very odd in that they are simultaneously eruptive and pulsating variables. Their main characteristic is a slow fading in light output by between 1 and 9 magnitudes that can last from a month to several hundred days. Their light curves as can be seen in Fig. 11.6 show a spectacular decline followed by a very slow recovery. It is thought that these stars are yellow F-type giants that are convective throughout and so may occasionally disgorge enormous amounts of carbon, which then condense at distances from the star, almost throwing a dark blanket over it. Eventual dispersal of this envelope results in the slow rise in brightness to its usual

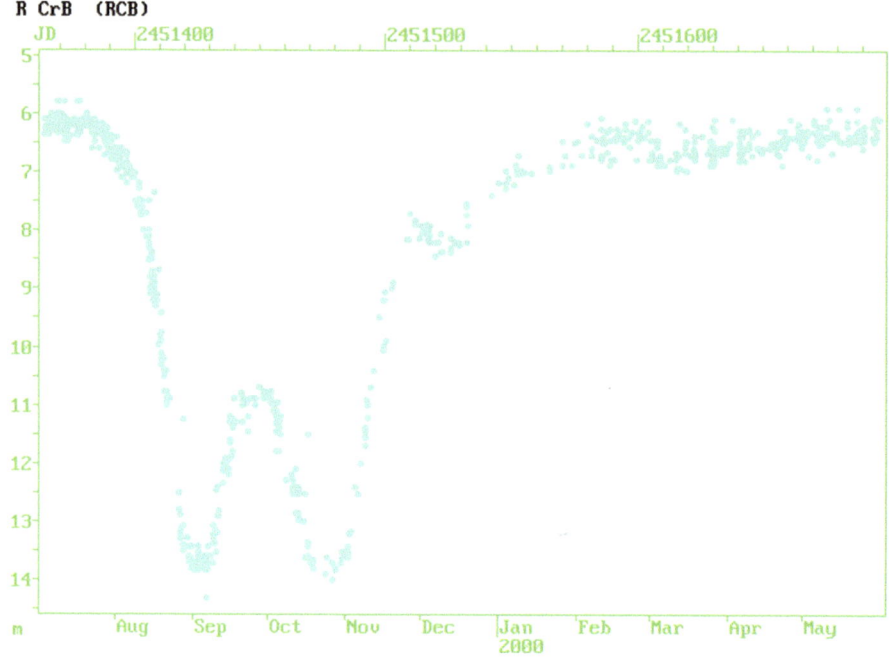

Fig. 11.6 R CrB light curve (Image from http://www.britastro.org/vss/.)

state. Other researchers think that such stars are probably the result of mergers between helium- and carbon-rich white dwarfs or even the result of a final He flash in post-AGB stars. No one is quite sure.

Whatever the mechanism, RCB stars are wonderful to follow. The only other bright one in its class is RY Sagittarii, and both stars hover around 6th magnitude for many years before an abrupt fade. Another star of similar type is the DY Persei variable.

DY Perseus

Rather than brightening quickly, these hydrogen-deficient stars show unpredictable fading events with slower declines and roughly symmetric recoveries. They almost appear to be like R Coronae Borealis-type stars, but their amplitude and magnitude of the declines is smaller than those of R Cor Bor stars.

Classical Novae

A star that suddenly gains in brightness and that has only been seen to erupt once is known as a classical nova. During outburst, the brightness of these so called novae (new stars) can increase by 6 to 19 magnitudes before fading back to their original level. The outbursts themselves can last anywhere from several days to years, but in general, the brighter the nova, the shorter its duration. In most of these type searches, digital SLR cameras and CCD cameras performing photometry are the norm.

Classical novae arise in a binary system where a star known as a white dwarf and a main sequence star orbit each other with a period generally less than 12 hours. Due to their close proximity and the extreme gravity of the white dwarf, hydrogen from the main sequence stars envelope is drawn into an accretion disk around the white dwarf and eventually deposited as a layer of hydrogen on the surface of the white dwarf. The pressure and temperature of this gas mounts over a period of time until it is hot enough to undergo hydrogen fusion. The entire gaseous envelope and part of the accretion disc is destroyed in the cataclysmic conversion of hydrogen to helium, and the star brightens considerably by up to 12 magnitudes. An illustration of the system can be seen in Fig. 11.7.

The energy released through this process pushes out the majority of the unburned hydrogen from the surface of the star in a shell of material moving at speeds of up to 1,500 km/s. In spite of the apparent violence, the amount

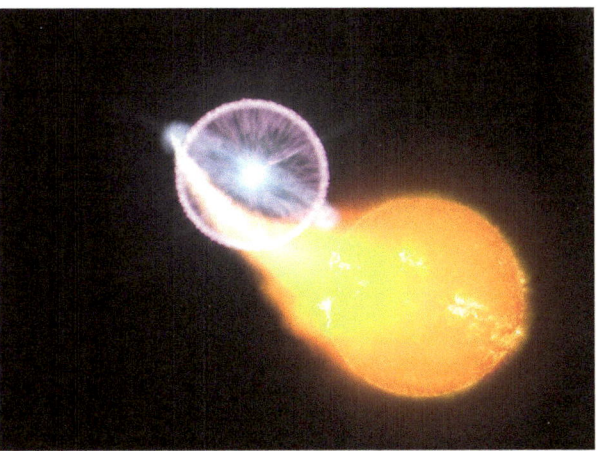

Fig. 11.7 Classical novae (Image from https://asd.gsfc.nasa.gov/blueshift/index. php/2017/09/05/shock-waves-power-an-exploding-star.)

of material ejected in novae is usually only about 1/10,000th of a solar mass, which is quite small relative to the mass of the white dwarf. It is considered that about 5% of the hydrogen envelope on the white dwarf undergoes fusion, and so there is enough material for recurrent explosions in the future.

In addition to the hydrogen-burning classical novae, another form of novae similar in mechanism but with different spectral characteristics has been observed. So called "helium novae" comprise a new category of novae event that lacks hydrogen lines in its spectrum. This could be due to the underlying white dwarf being primarily helium rich, and the novae mechanism could be caused by the explosion of a helium shell on a white dwarf. The first candidate helium nova to be observed was V445 Puppis in the year 2000, and since that time, another four similar types have been observed.

Notwithstanding their origins, classical novae are relatively rare, and in general it is not possible to predict when one will appear. Nevertheless, classical novae are spotted on a yearly basis in the Milky Way Galaxy, and novae patrol to discover them is confined to the disc of our galaxy, where the majority of stars reside. There are several subgroups of novae that astronomers now classify. These are:

NA

Fast novae, with a rapid brightness increase, followed by a brightness decline of 3 magnitudes to about 16% of the former brightness within 100 days of the outburst.

NB

Slow novae, with a 3 magnitudes decline in 150 days or more.

NC

Very slow novae, staying at maximum light for a decade or more, fading very slowly. However, such objects are very rare, and it is possible that NC-type novae are objects differing physically from normal novae. It has been suggested that they may be exhibiting proto-planetary nebulae like symptoms, but they also exhibit features in the spectra that mimic the very hot, bright Wolf-Rayet stars, so this is an ongoing field of study.

NR/RN

Recurrent novae, novae with two or more outbursts separated by 10 to 80 years, have been observed, and a typical example of this type is the star T Coronae Borealis.

Dwarf Novae

There are 200 billion stars in our Milky Way Galaxy, and of these over 55% are binary or multiple systems. It is small binary systems that are primarily responsible for the production of dwarf novae, a particular form of cataclysmic variable star. In dwarf novae, one star is a white dwarf – a collapsed star with the mass of the Sun in the volume of Earth – while the companion star is a red dwarf.

The red dwarf and the white dwarf orbit each other once every few hours; they are so close together that the average dwarf novae system would fit comfortably into the dimensions of our Sun. Due to this proximity, we cannot resolve the two stars. They appear on the sky as a point source. Most of what we know of dwarf novae comes from amateur study and the contribution of astrophysicists such as M. F. Walker, who deduced in the 1950s the binary nature of many cataclysmic variables, including dwarf novae.

The red star in a dwarf nova is so close to the white dwarf that it becomes tidally distorted, and gas is stripped off the red star and falls inward, towards the white dwarf, as can be seen from the illustration in Fig. 11.8. The gas does not plunge directly onto the surface of the white dwarf but swirls around it, producing an accretion disc. The gas in the disc spirals down towards the white dwarf, radiating its gravitational potential energy away as it goes. This energy becomes so intense that the accretion disc usually outshines both the red star and the white dwarf in visible light. The disc is stable up to a point, as the white dwarf does not have an intense enough magnetic field to disrupt the disc, and it is this characteristic, a lack of high magnetic induction (visible as a lack of Zeeman splitting of the spectral lines) that indicates that the stars are dwarf novae types. The gravitational energy release causes the outburst here, whereas in a classical nova, the accretion disc is usually destroyed as it undergoes thermonuclear detonation.

Collectively, dwarf novae are called U Geminorum-type variables. Rapid and small fluctuations in light output characterize such stars, but from time to time the brightness of a system increases rapidly by several magnitudes, and, after an interval of between several days to a month or more, they

Fig. 11.8 Dwarf nova system (Image from http://chandra.harvard.edu/photo/2001/v1494aql/.)

return to their original state. The interval between two consecutive outbursts for any given star may vary greatly, but every star is characterized by an interesting discovery, that the longer the cycle, the greater the amplitude. U Geminorum stars are often sources of X-ray emission, and it is thought that at least some of these systems are eclipsing binaries, possibly indicating that the primary minimum is caused by the eclipse of a hot spot that originates in the accretion disc from the infall of a gaseous stream from the K- or M-type star.

Dwarf novae outbursts are never going to be as spectacular as those of the rarer novae. (In outburst, dwarf novae are a mere factor of 6 to 100 brighter than in quiescence.) But their outbursts occur more often. As can be sensed from the foregoing, as a general rule of dwarf novae, the more frequent the outburst, the smaller the amplitude of outburst. A typical example of a dwarf nova is the star V1159 Orionis, which has outbursts once every four days, with an amplitude of about two magnitudes. At the other end of the scale is WZ Sagitae, which undergoes outbursts once every thirty years, and the outburst amplitude rivals that of a classical nova! Obviously, dwarf novae are very interesting beasts worthy of attention.

Dwarf novae therefore are a class of cataclysmic variable star that have multiple observed eruptions, but these only range in brightness from 2 to 5 magnitudes. They are generally faint stars that suddenly increase in brightness before fading back to their original magnitudes. And as can be seen from the above examples, the outburst intervals for each object are quasi-periodic, where their intervals can range from days to decades.

The lifetime of a dwarf novae outburst is typically from 2 to 20 days and are usually attributed to the release of gravitational energy resulting from an instability in the accretion disk or by sudden mass-transfers through the disk.

Within the dwarf novae class, there are three subtypes:

Z

Camelopardalis. Such stars exhibit standstills of about 0.7 magnitudes below the maximum brightness. Outbursts cease during these standstills for tens days to years.

SU

Ursa Majoris stars exhibit occasional super-outbursts that are typically 0.7 magnitudes, brighter than normal outbursts. The outburst lifetime in these cases is on the order of 5 times the lifetime of a normal outburst.

U Geminorum

These stars include all dwarf novae that are not members of the above subtypes.

One of the most interesting of the above star subgroups is the SU Ursa Majoris type. They show two distinct kinds of outburst: normal dwarf nova outbursts, and super-outbursts, which last 5 to 10 times longer and are slightly brighter than the usual dwarf nova outbursts.

These super-outbursts usually result from the stars having very short orbital periods, sometimes less than 2 hours. Some of the SU UMa stars with long outburst intervals show interesting brightening on the way back

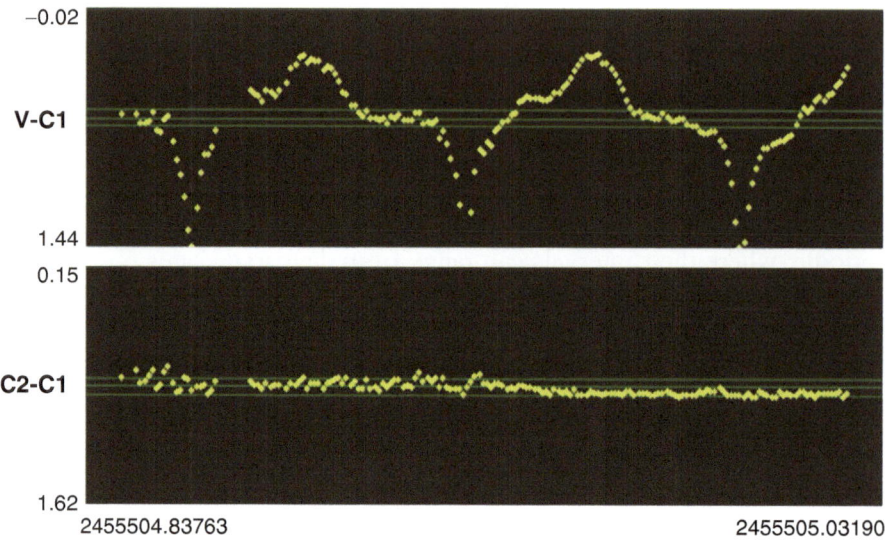

Fig. 11.9 Light curve with superhumps (Image from https://de.wikipedia.org/wiki/Superhump.)

to quiescence after a super-outburst. During a super-outburst, SU UMa stars reveal an additional modulation of the light curve, now called a superhump, caused by precession of the accretion disc around the white dwarf. Superhumps show up in the light curve as a modulation, with a period slightly longer than the orbital period, as can be seen from Fig. 11.9.

The following are also typical cataclysmic variables of the dwarf novae type and have their own designations in the AAVSO catalogue as progenitors of their type.

AM

AM Herculis (AM Her)-type variables are close red dwarf binary systems consisting of a dK- to dM-type dwarfs and a super strong magnetic white dwarf primary, in which the magnetic field not only prevents the formation of an accretion disc but also synchronizes the primary's rotation with its orbital period. They are a type of star known as polars due to the polarization of the light output by the huge magnetic fields. The variations of their light curves reveal a 4 to 5 magnitude amplitude in some AM Her stars, although the stars themselves are generally at 12th magnitude or fainter. The light curve of AM Herculis can be examined in Fig. 11.10.

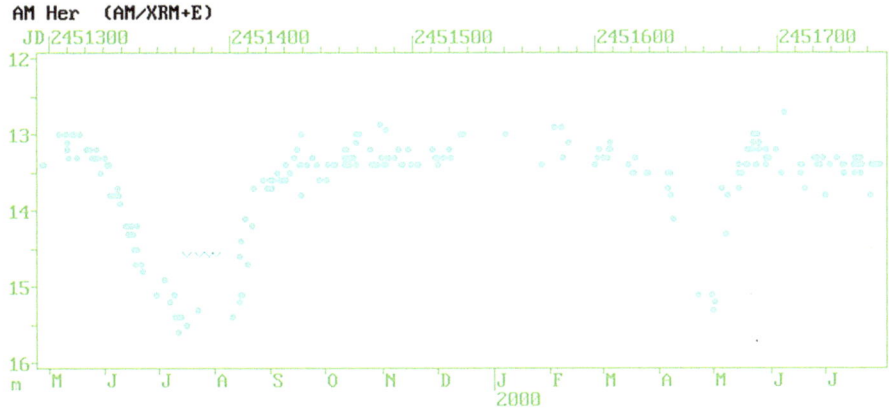

Fig. 11.10 AM Herculis light curve (Image from http://www.britastro.org/vss/.)

CBSS

The initials above stand for close binary super-soft source, which refers to the soft X-rays that are being produced in such high luminosity by accretion of hydrogen onto the surface of a white dwarf, which is undergoing a steady nuclear fusion. The material is driven off the primary star by a steady stellar wind, and thus the accretion is a slow and measured process. Typically, their orbital periods range from 0.15 to 4 days. It has been discovered that if the orbital period is less than 6 hours this mechanism does not work. In some CBSS stars their light curves show unusual patterns that may indicate the presence of jets of gas, probably from the polar regions, where magnetic confinement can produce some outflow.

A subgroup of these faint variable stars is the V Sagittae-type stars. They belong to the CBSS class but are not detected as super-soft X-ray emitters. In V Sge stars, their spectral lines are usually stronger and broader and resemble the following type.

DQ

These stars are magnetic cataclysmic variables with a red dwarf secondary and a white dwarf primary component that generates a magnetic field weaker than the field associated with AM Herculis stars and that is not strong enough to synchronize the orbits of the rotating white dwarf with the orbital period of the system. They are named after the progenitor star DQ Herculis and are also known in variable star circles as IP type stars or inter-mediate polars.

IBWD

This stands for interacting binary white dwarfs. These are close binary systems with very short periods of between 5 to 70 minutes, and their light fluctuations are very small, typically fractions of a magnitude, and thus are best covered photometrically. Occasionally, these stars are known as AM Canes Venaticorum stars or helium dwarf novae because they lack hydrogen lines in their spectra.

Supernovae

Supernovae are very rare explosions of supergiant stars. Their rarity precludes most amateurs from engaging in hunting for such objects, but within the last 25 years improved techniques, organized supernova patrols and the ability to image distant galaxies quickly and process the results has led to the rise of the dedicated amateur in this field.

Do not underestimate how difficult this task can be. The observer will spend many hours imaging hundreds or even thousands of galaxies just to find one supernova. You must have access to excellent sky charts, join a dedicated team of other observers who will give tips and advice and can check any events that you may have discovered. It has been estimated by some of these dedicated observers that for every supernova potential there may be between 20 or 30 observations that need to be checked. And of course there is also the weather to contend with! Nevertheless, supernovae do occur in external galaxies on a regular basis. Figure 11.11 shows a Type Ia supernovae in Messier 82 imaged by the author in January 2014.

Checking your images against existing databases is absolutely crucial. There are several to choose from, such as the Supernovae Search Tool from the Puckett Observatory Supernova Search Team, which works with the commonly available Maxim DL and can be found at https://diffractionlimited. com/help/maximdl/MaxIm-DL.htm#Supernova_Search_Tool.htm.

Additionally, there is the European Southern Observatory's Deep Sky Survey, which can be found via http://archive.eso.org/dss/dss. There is also the Sloan Digital Sky Survey and the Palomar Observatory Sky Survey. Many of these tools and others are accessible online.

Finding an apparently new object on a patrol image is far from being a guarantee that a supernova has been discovered. This is true even in cases where the object is embedded well within a galaxy's spiral arms, and for very experienced searchers, it is intuition that tells you that it is obviously a supernova. Below are some suspects to eliminate from your inquiries.

Fig. 11.11 Supernovae in Messier 82 (Image by the author.)

Asteroids are a very common cause of false alarms. These appear on images any time that you are patrolling close to the ecliptic and even on occasions when you think you are safely far enough away from it. Beware of these interlopers! The best way to see if the object is an asteroid is to check with the supernova candidate Minor Planet Checker (MPC). If the MPC software does not identify it, then multiple images will be required over a period of hours; the objective being to have its movement give itself away. If it does turn out to be an unknown asteroid, of course, you have still made a discovery but of a different kind. However, even then if a newly discovered asteroid has not yet reached its first opposition the MPC may draw a blank.

The CCD chip may be the subject of hits from cosmic rays or other ionizing radiation. Such impacts appear as a very sharp spot similar to a hot pixel or an object with an unusual shape. In cases where they may resemble a star the only option is to re-image the galaxy.

Don't forget that variable stars can appear on the chip, especially such stars as Cepheids, whose lower magnitudes might be below the sensitivity of the chip. Of course, there is no completely accurate test for variable stars. It is possible that the digitized sky surveys currently online may have some knowledge of their presence.

Anyone starting off in supernovae patrol would be well advised to work alongside someone with a great deal of experience. The British Astronomical Association and many other astronomical associations worldwide have supernova patrols and officers dedicated to helping someone get the best from their searching. We all have to start somewhere, and whatever your variable star observing experience, this type of patrolling is a serious undertaking.

When a suspect is first reported there are certain basic things that must be known about it. The time and date of supposed discovery is required, the position in the sky to approximately one arcsecond, its approximate magnitude and the offsets (in arcseconds) from the galaxy's core.

These offsets help whoever will be taking any further images to find the supernova. Your report should find its way to the Central Bureau for Astronomical Telegrams, who will want to know the limiting magnitude of your discovery image and have access to some good quality archived images where the object is not visible. It will be necessary to show that you have also checked the European Southern Observatory's Deep Sky Survey red and blue plates. It is highly important to give the time and date that each image was taken and estimate each plate's limiting magnitude.

Supernovae, then, are stars that increase their brightness by 20 magnitudes and more before fading slowly. The expansion velocities of supernova envelopes are in the thousands of km/s, and they are the brightest types of cataclysmic variable and also among the rarest. Following their light curves, supernovae are subdivided into Types I and II. Rather confusingly Type II supernovae have traditionally been thought to be massive stars, while Type I are white dwarfs. However, follow the descriptions below for a better analysis. Figure 11.12 gives some indication of the light curves and their differences.

Type I Supernovae

These can be a rather confusing mishmash of types, but let us start with the classical white dwarf collapse as it accretes mass and goes over the Chandrasekhar limit of 1.4 solar masses.

Such stars are undergoing the same type of activity, and so all explode with a peak absolute magnitude of −19.3. There is generally no remnant, though it is possible that a neutron star can be formed in some Type Ia explosions. As can be seen from Fig. 11.12, a type Ia supernova rapidly diminishes in brightness over 20 to 30 days following maximum light. Thereafter, the brightness decreases by approximately 0.1 mag. per day, before the fading begins to tail off to reach a value of 0.014/day.

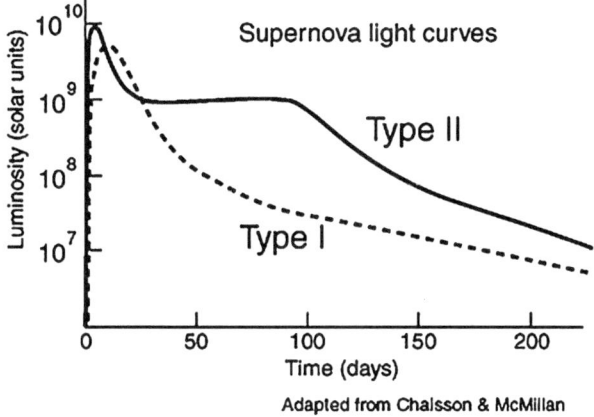

Fig. 11.12 SN light curves (Image from http://hyperphysics.phy-astr.gsu.edu/hbase/Astro/snovcn.html.)

The other Type Ia supernovae is the type Iax, in which a carbon- and oxygen-rich white dwarf accretes matter from a helium star that lost its outer hydrogen envelope during its late evolution peak. Its absolute magnitude varies between −14.2 and −18.9. It is, though, possible that the white dwarf is not completely destroyed.

These sum up the supernovae from white dwarf stars. The following are distinctly different.

Type II Supernovae (and Odd Type I's)

These types are typified by massive stars with lines of hydrogen and other elements in their spectra. The expanding envelope consists mainly of H and He, and their light curves show greater diversity than those of Type I supernovae. From Fig. 11.2 it can be seen that they fade slower, with a defined hump before tailing off after 40 to100 days or more.

SN 1B are intermediate mass Wolf Rayet star progenitors with a predominance of nitrogen in their atmosphere before eruption. They are commonly called stripped core-collapse supernovae. Both these and their close cousins, SN Ic supernovae, have Wolf Rayet stars as originators with the spectral difference that Type Ic have carbon or oxygen rich progenitor stars.

SN IIA are a mixture of stars that originate in carbon- or oxygen-rich white dwarf stars with a main sequence companion of intermediate mass of between 6 or 7 solar masses. They are generally characterized by circumstellar shells of materials stripped from the companion and reach a peak magnitude of around −18.

At the time of writing there are some conflicting ideas about type SN 1B. Are they intermediate mass Wolf Rayet progenitors? Massive binary stars? Type IIb are known by their rapidly declining light curve.

SN IID are typified by the occurrence of strong wind ejection episodes shortly before the explosion. Flattening in the light curve at later stages is due to the interaction between the supernova ejecta and the existing circumstellar material.

SN IIN are massive stars that are typically LBV progenitors. The explosion raises a great deal of envelope material, and their absolute magnitudes can vary between −17 and −20 magnitude.

SN II-P are the classical relatively low mass red supergiants that cook elements up to iron and then undergo core collapse They are designated II- P as they have an extended "plateau" to the light curve before they fade away. They have observed absolute magnitudes between −16 and −18.

It is unlikely that the average amateur would have access to the necessary spectroscopic equipment to discern between the different types of supernova. However, just finding a supernova is an uncommon event and very worthy of note! Naturally, supernovae are rare due to the dearth of large stars and due to the fact that the rapid expansion will dissipate the gas. Therefore, even seeing a SN is a bonus, as these can be very challenging especially from urban areas.

For those observers who wish to follow these unusual stars, the BAA variable star section has an exhaustive list of variables that can be followed by amateur observers. These important fields of work are detailed at the following URL: http://www.britastro.org/vss/.

Chapter 12

Unusual Variables – X Ray and Visual GROs

Observers who are looking for a particular challenge may be taken by the fact that extending into the short wavelengths of the electromagnetic spectrum, yet still visually observable in part, are a number of objects collectively known as X-ray variables and gamma ray objects. Naturally, these will be relatively rare, as they generally involve high mass stars or objects that are the subject of ongoing stellar research. Nevertheless, a light curve may be obtained photometrically from such objects if one has patience and determination.

Such variable stars are highly unusual in that they could be late or early type stars depending on the processes behind the source. Analysis by the SWIFT gamma ray satellite and large ground telescopes combined with the Hubble Space Telescope found a number of transient gamma ray flares or bursts (GRB). About 40% of them have a redshift (z) greater than 1, meaning that they were at cosmological distances and are possibly supernova flares in remote galaxies. However, what are the other 60%? In all probability, astronomers were looking at not just a single class of object, and so the search was expanded and turned up the fact that some main sequence stars of M spectral class were responsible for some flares. It has subsequently been discovered that faint stars (less than 8th magnitude visually) of G, K, and M spectral classes could be gamma-ray burst sources due to their flare activity.

Alongside these energetic events are the more low key, but highly interesting, X-ray variables, which, until recent advances in spaceborne

© Springer Nature Switzerland AG 2018   175
M. Griffiths, *Observer's Guide to Variable Stars*, The Patrick Moore
Practical Astronomy Series, https://doi.org/10.1007/978-3-030-00904-5_12

technology, were all but unknown to the ordinary observer and turned up some surprising results. For example, the prototype star after which all Cepheids are named, δ Cephei is an X-ray variable. Data recently returned from the Chandra X-ray Observatory, combined with previous X-ray measures secured with the XMM-Newton X-ray satellite, have shown that δ Cephei has X-ray variations occurring in accord with the supergiant star's 5.4-day pulsation period. X-rays are observed at all phases of the star's pulsations, but sharply rise by approximately 400% near the times when the star swells to its maximum diameter of about 45 times that of the Sun.

X-ray activity seems to be a large part of the evolution of large stars and can occur across a large section of their stellar types.

Main Types of X-ray Variable Stars

HMXB

This is an acronym for high mass X-ray binaries. Such systems involve a massive star, usually a supergiant of O or B star type and a compact object that could be a neutron star or a white dwarf companion. In some cases, the X-rays are known to originate in the accretion disc around a black hole, so these are very exciting stars! The process that emits X-rays is caused when a fraction of the stellar wind of the normal star is captured by the compact object and produces X-rays as it falls onto it or onto an accretion disc that's surrounding it. In addition, a close companion could be stealing mass from the large primar, and the X-rays come from either the hot disc as it falls into the compact object or from a hot spot as the materials impact upon the accretion disc. Fig. 12.1 illustrates the type of process at work in most X-ray binaries.

In X-ray binaries, the subtypes indicate which kind of behavior the binary displays, like the XB stars that undergo X-ray bursts or the XN stars that have large amplitude outbursts in the visual range. There is also variability due to reflection from the hot surfaces of the giant stars, and these are known as XR variables. There is also a class of variables known as XP in which the compact object is a pulsar. All in all, these are very interesting objects, and their production processes are similar to the low mass X-ray binaries that follow here.

Fig. 12.1 X-ray binary and accretion disc (Image from https://www.quora.com/ Is-it-possible-for-the-accretion-disk-of-a-black-hole-to-emit-as-much-light-as-our-sun-but-not-significantly-more-harmful-radiation-than-it.)

LMXB

The low mass X-ray binaries are wonderful systems where one of the components is either a black hole or a neutron star. The donor star usually fills its Roche lobe and therefore transfers mass to the compact object. The donor can be a normal dwarf, a white dwarf, or an evolved star such as red giant. X-rays are emitted as the mass falls onto the compact object or onto an accretion disk that's surrounding it. The X-ray emission is incident upon the atmosphere of the cooler companion of the compact object and is reradiated in the form of optical high-temperature radiation (reflection effect), thus making that area of the cooler companion's surface an earlier spectral type.

In all X-ray binaries, the subtypes indicate which kind of behavior the binary displays, as we have seen with the XB, XR and XP variables above. The X merely stands for an X-ray source.

XB Variables

These stars are X-ray bursters that are close binary systems undergoing X-ray and optical bursts with durations from several seconds to ten minutes and with small amplitudes of about 0.1 magnitudes in the visual range. Their light curves look as if the star is sputtering as material rains onto it.

XJ

XJ variables are X-ray binaries that are characterized by the presence of relativistic jets evident at X-ray and radio wavelengths as well as in the optical. In a spectroscope, these stars show emission line components with relativistic velocities. They are generally thought to be supergiant stars with black hole secondary components, and the visual range of variability is between 0.1 and 0.3 magnitudes.

XN

Systems with neutron stars also show variability in the visual; such X-ray systems occasionally rapidly increase in brightness by about a magnitude over short periods of a few minutes.

XP

These are X-ray pulsar systems and a combination of a primary component, usually a supergiant star of O or B type but in close orbit with its small companion, dragging the giant out into an elliptical shape. In such stars the light variability is mainly caused by the ellipsoidal primary component's rotation. Observers have recorded periods of light changes between 1 and 10 days, but the amplitude of the light curve is no more than 0.2 magnitudes in the main.

XPR

These are X-ray pulsar variables that feature the reflection effect. The mean visual light output of the system is brightest when the primary component is irradiated by X-rays and they can be very varied in amplitude, with differences of between 2 to 3 magnitudes in the visual range recorded for some of these stars. HZ Herculis is a typical example of this class and usually fluctuates between 13 and 15th magnitude. Fig. 12.2 shows the light curve for this object.

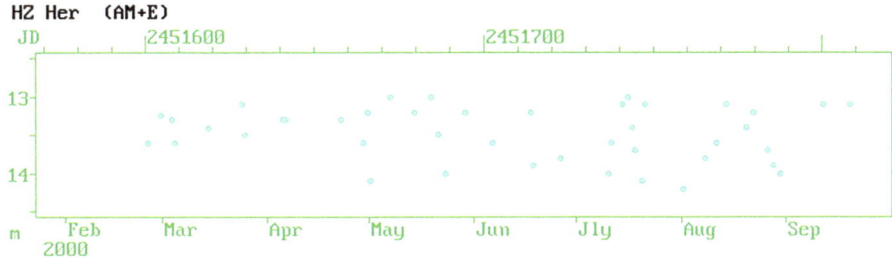

Fig. 12.2 HZ Herculis (Image from http://www.britastro.org/vss/.)

ZZ

In addition to the supergiant stars above, there are some low mass stars that, due to flare activity, also radiate in X-rays. These are commonly known as ZZ Ceti variables. Such stars are non-radially pulsating white dwarfs that change their brightnesses with periods ranging from 30 seconds up to 25 minutes and have small amplitudes that range from 0.001 to 0.2 magnitudes in the visual, revealing that these stars are small, and the processes underpinning their variability are fleeting. Occasionally, flares of up to 1 magnitude are sometimes observed, and it has been postulated that such stars may be binaries with UV Ceti-type companions all adding to the mix.

Variables of a different type that have even shorter main wavelengths are known as optical gamma ray emitters. Such objects are not well studied and are in the main extragalactic objects that show fleeting visual counterparts. Known as OGRE's, these are difficult objects to observe, and there is no knowing when a burst may arise from such an object. Back in the 1980's one such source was noticed in the constellation of Taurus, which appeared to repeat occasionally.

Gamma ray bursts may originate in nova or supernovae explosions as well as in high intensity interactions in binary stars, or in materials swirling into black holes. In some cases of high energy X-rays and gamma rays, the sources have been shown to be extended objects such as dust haloes around supergiant stars. A progenitor type is the so-called variable star BL Lacertae, which was eventually discovered to be a quasar at a great distance from us.

For the average observer, gamma ray astronomy is going to be way out of your league, and the visual components to such objects are generally remote, uncertain of period and type and relatively rare. Anyone who still wishes to look for any potential object that could be worth the time and investigation can refer to either the CHANDRA or Compton space telescope observatory catalogue or to specialist publications.

Chapter 13

Variable Star Associations

Astronomy may seem like a rather solitary occupation – up all night with the world asleep all around you while you are looking intently at various objects with a fascination all of their own. Variable stars demand time, patience and long-term study and seem on the surface to be an anti-social sort of pursuit. However, the loneliness of the long distance observational astronomer can be assuaged by interaction with several groups that tie astronomers together in bonds of camaraderie and friendship and in which their endeavors are valued and their results published.

The study of variable stars lagged some distance behind Solar System, positional, astronomy double star measurement and deep sky research until the middle part of the 19th century. Then, following F. W. A. Argelander's pioneering work in the 1840s, there was a striking increase in variable star research, particularly in Europe. The transformation was to such an extent that in the second half of the 19th century there were three attempts at forming variable star associations within Great Britain alone. The first, in 1863, was the Association for the Systematic Observation of Variable Stars (ASOVS), which never got off the ground. The second in 1883 was the Liverpool Astronomical Society Variable Star Section, which had somewhat limited achievements. The third, launched in 1890, was the BAA VSS, which was eventually both a resounding and lasting success.

Today, there are two main repositories for variable star work that, between them, have brought variable stars and variable star observers together and become beacons of research and scholarly activity. The oldest

© Springer Nature Switzerland AG 2018
M. Griffiths, *Observer's Guide to Variable Stars*, The Patrick Moore
Practical Astronomy Series, https://doi.org/10.1007/978-3-030-00904-5_13

of these institutions is the British Astronomical Association (BAA) and its variable star section. The Second is the American Association of Variable Star Observers (AAVSO).

The BAA VSS

The Variable Star Section (VSS) of the BAA was formed in 1890, the same year the BAA was founded, with the aim of collecting and analyzing observations of variable stars that could be then compared, reduced or sent out for further professional analysis. Since its inception, the VSS has been run by a small group of officers who deal with various aspects but meet at regular intervals to discuss and decide future plans and policies. Feedback to BAA members is through the VSS Circulars published four times a year, commonly through the *BAA Journal*. The journal itself provides a peer-reviewed forum for broadcasting of variable star results.

Up until 1934, VSS records and observations were published as *BAA Memoirs*. Most records from 1890 right up until 1990 were kept as handwritten original reports and chronological lists. Such records serve as a long history of observations and a historical resource that can be checked and cross indexed, giving instant access to variable star observations across a century of time and a host of variable star types. The whole wealth of materials has also been computerized, and the section is a very professional and research-led body.

The BAA variable star section has hundreds of international observers on its books, and there are overlaps with their observations and those of the BAA's American counterpart.

The AAVSO

Since its founding in 1911 by the astronomer William Olcott, the American Association of Variable Star Observers (AAVSO) has coordinated, collected, evaluated, analyzed, published, and archived variable star observations made largely by amateur astronomers and makes the records available to professional astronomers, researchers, and educators. These records establish light curves depicting the variations in the brightness of a star over time.

Since professional astronomers do not have the time or the resources to monitor every variable star, astronomy is one of the few sciences where amateurs can make genuine contributions to scientific research. During

2011, the 100th year of the AAVSO's existence, the 20-millionth variable star observation was received into the database. The AAVSO International Database currently stores over 35 million observations. The organization receives nearly 1,000,000 observations annually from around 2,000 professional and amateur observers and is quoted regularly in scientific journals.

The AAVSO is also very active in education and public outreach. It routinely holds training workshops for citizen science and publishes papers with amateurs as coauthors. In the 1990s, the AAVSO developed the Hands-On Astrophysics curriculum, now known as Variable Star Astronomy. In 2009, the AAVSO was awarded a three-year grant from the NSF to run Citizen Sky, a pro-am collaboration project examining the 2009–2011 eclipse of the star Epsilon Aurigae, which has yielded some fascinating results.

The AAVSO headquarters were originally located at the residence of its founder, William T. Olcott in Norwich, Connecticut. After AAVSO's incorporation in 1918 it de facto moved to Harvard College Observatory, which later officially provided an office as the AAVSO headquarters (1931–1953). After that time it moved around Cambridge before purchasing their first building in 1985 – The Clinton B. Ford Astronomical Data and Research Center. In 2007, the AAVSO purchased and moved into the recently vacated premises of *Sky & Telescope* magazine, and from here it coordinates variable star observing activities across the globe.

The AAVSO currently receives variable star brightness estimates from about 1,000 amateur astronomers per year. No observations, if they are of sufficient quality, are ever turned down. Some variable stars are bright enough to be seen with the unaided eye, while others require high-tech equipment. To encourage observations, the AAVSO also has a network of robotic telescopes available to members free of charge.

Because some variable stars are unpredictable and can change their brightness over long time scales, it is not practical for professional astronomers to watch them every night. Thus, amateurs have been, and are being, recruited to keep tabs on these stars on behalf of professionals. Between the BAA and the AAVSO, there are over 40 million observations of variable stars available for perusal by scientists and amateurs alike.

It is easy to underestimate how valuable this data is. However, consider that the work of these organizations represent the works of countless people who have left records and observations that inspire new generations. The database itself spans many generations and includes data that cannot be reproduced elsewhere. If an astronomer wants to know the history of a particular star, they come to the AAVSO or the BAA.

If any observer wants further information about any aspect of variable star observing, then it is recommended going to either of the two following websites:

https://www.aavso.org
http://www.britastro.org/vss/

Internationally there are also variable star observing sections or societies that also cater to the amateur astronomer no matter what his or her experience. In most continents there are variable star associations that provide help, data, meetings and activities for all comers. Their websites can be found by checking out these websites:

Australia: https://www.assa.org.au/sig/variables/
Belgium: http://www.vvs.be/werkgroepen/werkgroep-veranderlijke-sterren
Canada: https://www.rasc.ca/variables/index.shtml
Czech Republic: http://var2.astro.cz/index.php
France: http://cdsarc.u-strasbg.fr/afoev/index.htx
Germany: http://www.bav-astro.eu/index.php?sprache=en
Japan: http://vsolj.cetus-net.org
Netherlands: http://www.veranderlijkesterren.info
New Zealand: https://www.variablestarssouth.org
Sweden: http://www.saaf.se/sektioner.php

Chapter 14

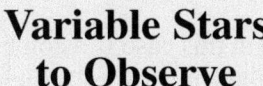

Variable Stars to Observe

The following section contains a season by season account of various variable stars that the reader may find of interest. We have tried to include information from a variety of sources so that any observer can go from faint flare stars or recurrent novae to long-period variables and enjoy many rewarding activities charting the light curves of each star.

The list is not exhaustive, merely a catalogue of objects that the author finds of interest. No doubt other observers will have their favorites, and of course none of these need to be followed by more experienced observers, as they probably have targets of interest themselves that command their attention.

Nevertheless, do enjoy this selection of stars and see how many of them appeal to you. There may be some overlap with the list of binocular variables, but such overlap does not detract from the intrinsic interest of each star.

© Springer Nature Switzerland AG 2018   185
M. Griffiths, *Observer's Guide to Variable Stars*, The Patrick Moore
Practical Astronomy Series, https://doi.org/10.1007/978-3-030-00904-5_14

EG ANDROMEDAE
Type: ZAND
Magnitude range: 7.1 to 7.8
Period: –
RA: 00.41.53   Dec: 40.24.4

This is a symbiotic star that in this case is a combination of a cool red giant and a white dwarf in which the white dwarf's accretion disc is either tangled with the red giant or is responsible for occulting the larger star. Such symbiotic stars occasionally show minor outbursts that make the star flare for a brief period. EG Andromedae is very close to the galaxy Messier 31, and the star shows irregular fluctuations in its spectra and light output, making it vary by up to 0.2 magnitudes over a semi-regular period of 40 days or so. It is an object worth keeping an eye on, and being at magnitude 7 in the main, it is visible in binoculars.

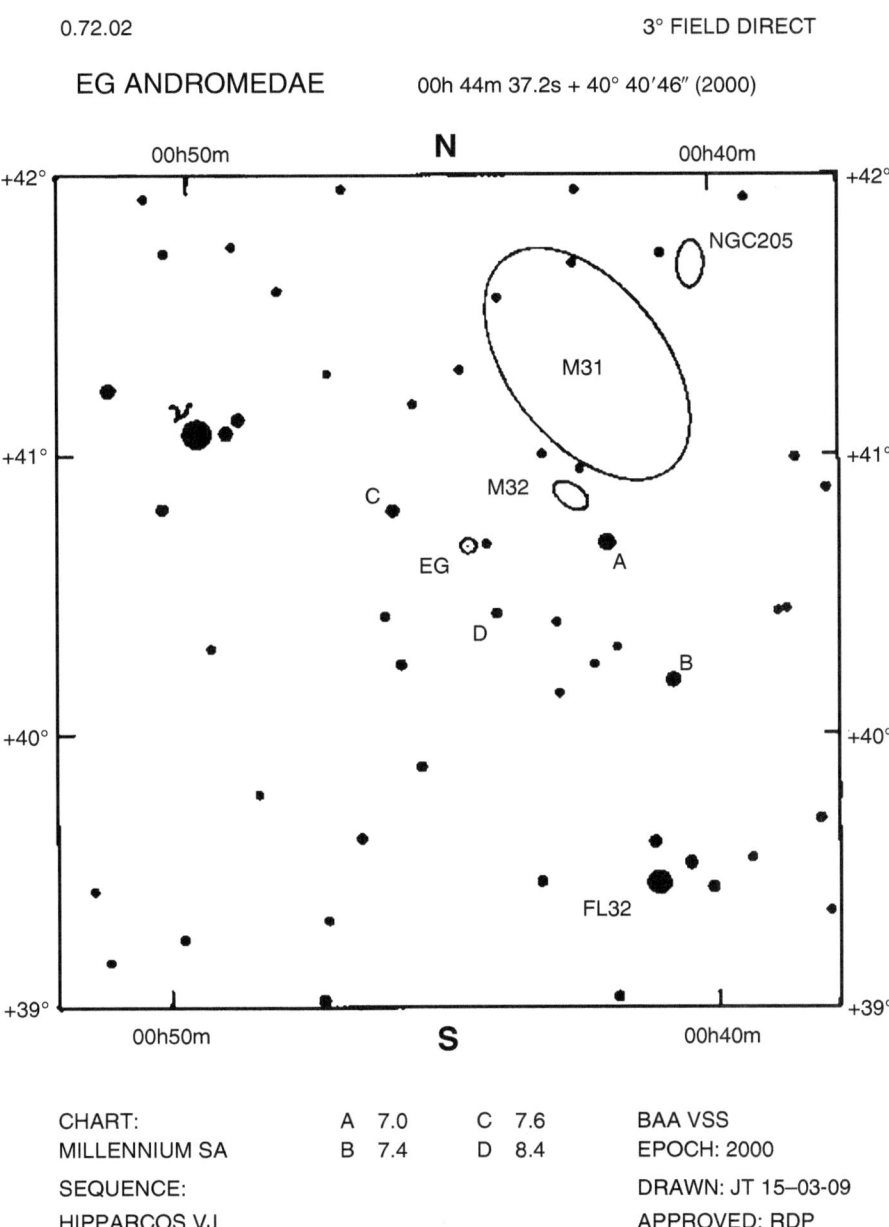

EG ANDROMEDAE   00h 44m 37.2s + 40° 40′46″ (2000)

CHART:	A 7.0	C 7.6	BAA VSS
MILLENNIUM SA	B 7.4	D 8.4	EPOCH: 2000
SEQUENCE:			DRAWN: JT 15–03-09
HIPPARCOS VJ			APPROVED: RDP

Fig. 14.1 EG Andromeda

(Finder chart courtesy of the British Astronomical Association Variable Star Section)

RX ANDROMEDAE
Type: UGZ
Magnitude range: 10.3 to 15.1
Period: 14d
RA: 01.04.35   Dec: 41.17.58

Although a very faint star at its usual magnitude, this cataclysmic variable star has a period of just two weeks and can be seen rising quickly from magnitude 15.1 up to magnitude 10 in a matter of moments before fading over a few hours. Typical of its class of Z Camelopardalis variables, these dwarf novae are challenging objects in that their maximum are quick, and there is no guarantee that even if there is an alert, by the time darkness may fall at your location, it may have returned to minimum brightness. Nevertheless, experienced observers looking for a challenge may keep an eye on this dwarf nova.

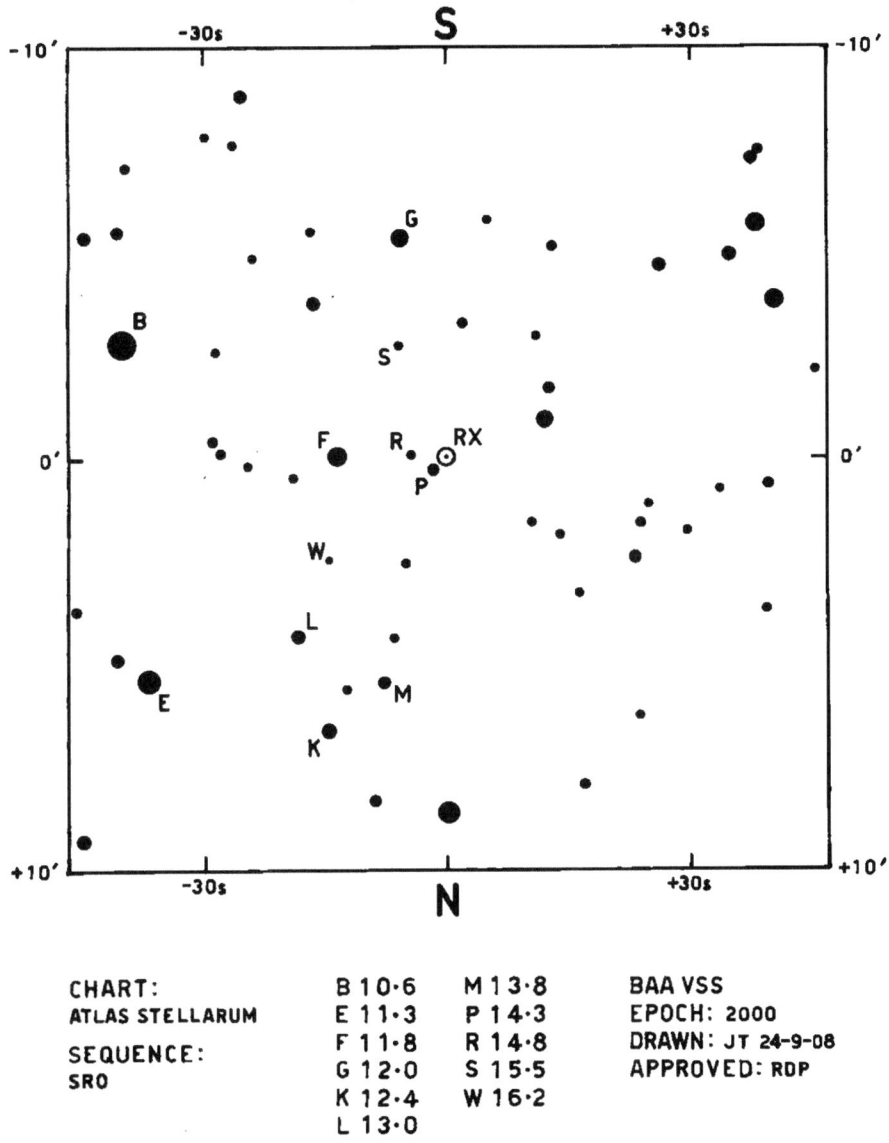

001·04   20' FIELD INVERTED

RX ANDROMEDAE   01h 04m 35·5s   +41°17'58" (2000)

CHART:	B 10·6	M 13·8	BAA VSS
ATLAS STELLARUM	E 11·3	P 14·3	EPOCH: 2000
SEQUENCE:	F 11·8	R 14·8	DRAWN: JT 24-9-08
SRO	G 12·0	S 15·5	APPROVED: RDP
	K 12·4	W 16·2	
	L 13·0		

Fig. 14.2 RX Andromeda

(Finder chart courtesy of the British Astronomical Association Variable Star Section)

W ANDROMEDAE
Type: M
Magnitude range: 6.7 to 14.6
Period: 397d
RA: 02.17.32   Dec: 44.18.17

This beautiful long-period variable star is a red giant of S stellar class, so is a rare "carbon" star that has been observed for over a century. A modest telescope will allow the observer to follow it from peak to minimum brightness, and due to its class it is intensely red and easy to see in the field of view. Like many of its type, the peak output of the star is in infrared and studies of the star show that throughout its variable cycle, the bolometric magnitude does not change in accordance with the visual output.

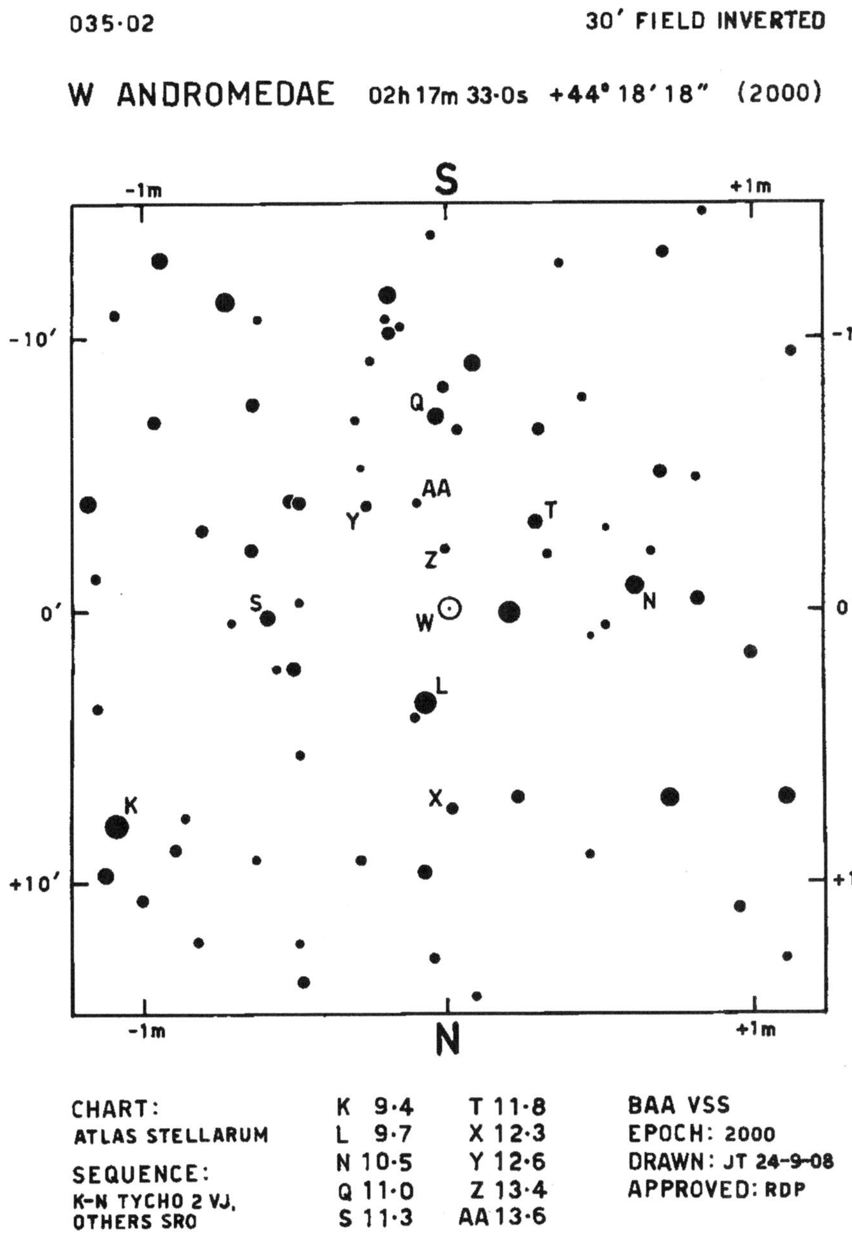
035·02
30′ FIELD INVERTED
W ANDROMEDAE 02h 17m 33·0s +44° 18′ 18″ (2000)
S
N
-1m
+1m
-10′
0′
+10′
Q
AA
Y
T
Z
N
S
W
L
K
X
CHART:
ATLAS STELLARUM
SEQUENCE:
K–N TYCHO 2 VJ,
OTHERS SRO
K 9·4
L 9·7
N 10·5
Q 11·0
S 11·3
T 11·8
X 12·3
Y 12·6
Z 13·4
AA 13·6
BAA VSS
EPOCH: 2000
DRAWN: JT 24-9-08
APPROVED: RDP

Fig. 14.3 W Andromedae

(Finder chart courtesy of the British Astronomical Association Variable Star Section)

VY AQUARII
Type: UGSU
Magnitude range: 8 to 16.6
Period: –
RA: 21.09.28   Dec: –09.01.50

The variable star VY Aquarii has been regarded as an outstanding member of the class of recurrent novae because it has erupted several times with amplitudes greater than 8th magnitude, which is higher than many other dwarf novae. There appears to be a 120-minute resonance to the star that may indicate the orbital period of the companion in this dwarf system.

This star is a favorite with experienced variable star observers, as it is faint, but photometry of its light curve reveals the presence of so called superhumps with a period of 92.7 minutes. The timings also drift somewhat, and this process is thought to be due to precession of the accretion disc surrounding the white dwarf. VY Aquarii in outburst shows remarkable activity, and the minimum and maximum light output are never regular and show dissimilarities in each outburst, hinting at the chaotic nature of the close binary stars and the disc that surrounds one. This is a faint star, but at peak outburst it is evident in the field of view of a small telescope.

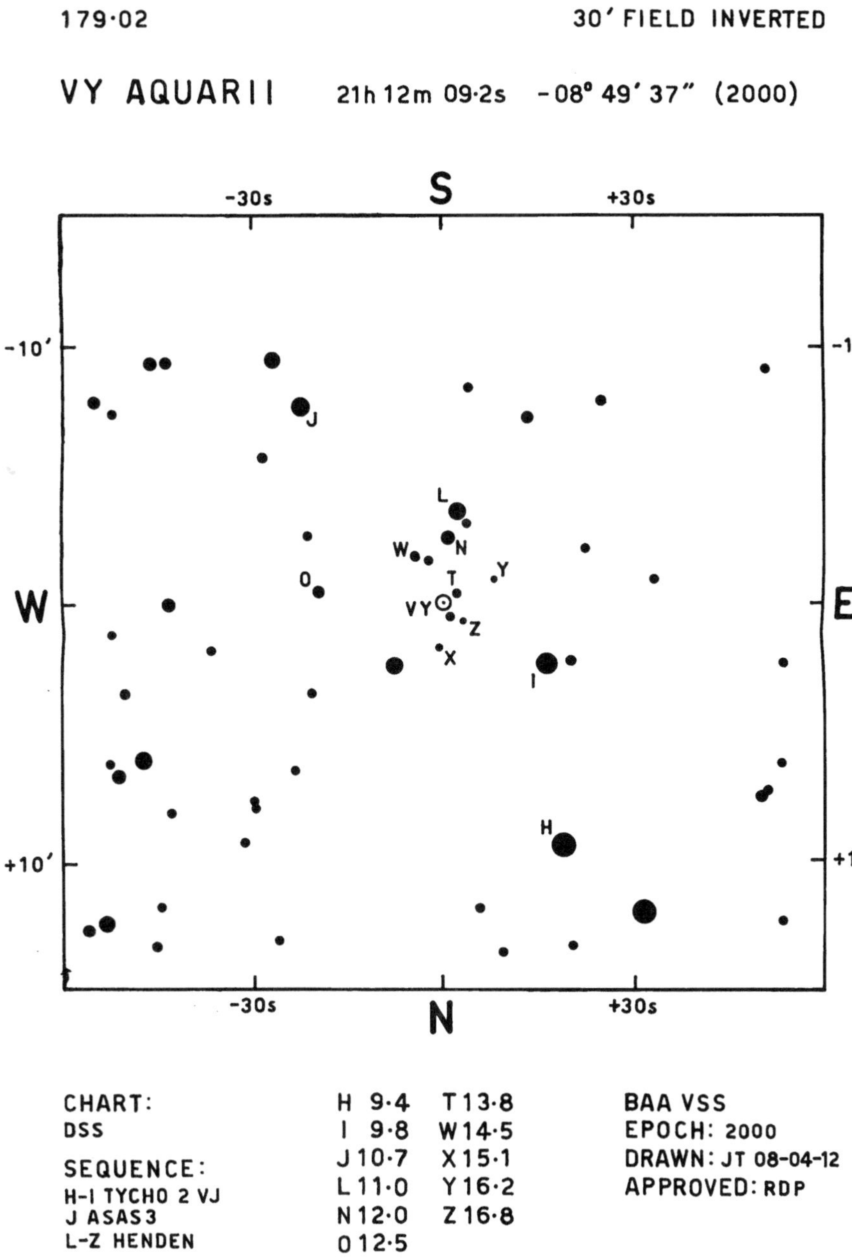

179·02   30' FIELD INVERTED

VY AQUARII   21h 12m 09·2s   −08° 49′ 37″ (2000)

Fig. 14.4 VY Aquarii

(Finder chart courtesy of the British Astronomical Association Variable Star Section)

OMICRON CETI
Type: M
Magnitude range: 2 to 10.1
Period: 332d
RA: 02.16.49   Dec: −03.12.2

Omicron Ceti, or Mira, is one of the most studied long-period variable stars in the heavens. Discovered over four centuries ago, its slow rise and fade to obscurity have fascinated astronomers. The star is an asymptotic red giant in the final stages of its life, and studies by Hubble and GALEX have shown a disfigured, out of shape star trailing a ghostly, comet-like wreath of gas as it moves in proper motion across the sky. Mira is well on the way to becoming a proto-planetary nebulae. In the meantime, watch the majestic rise and fall of light in this superb little object.

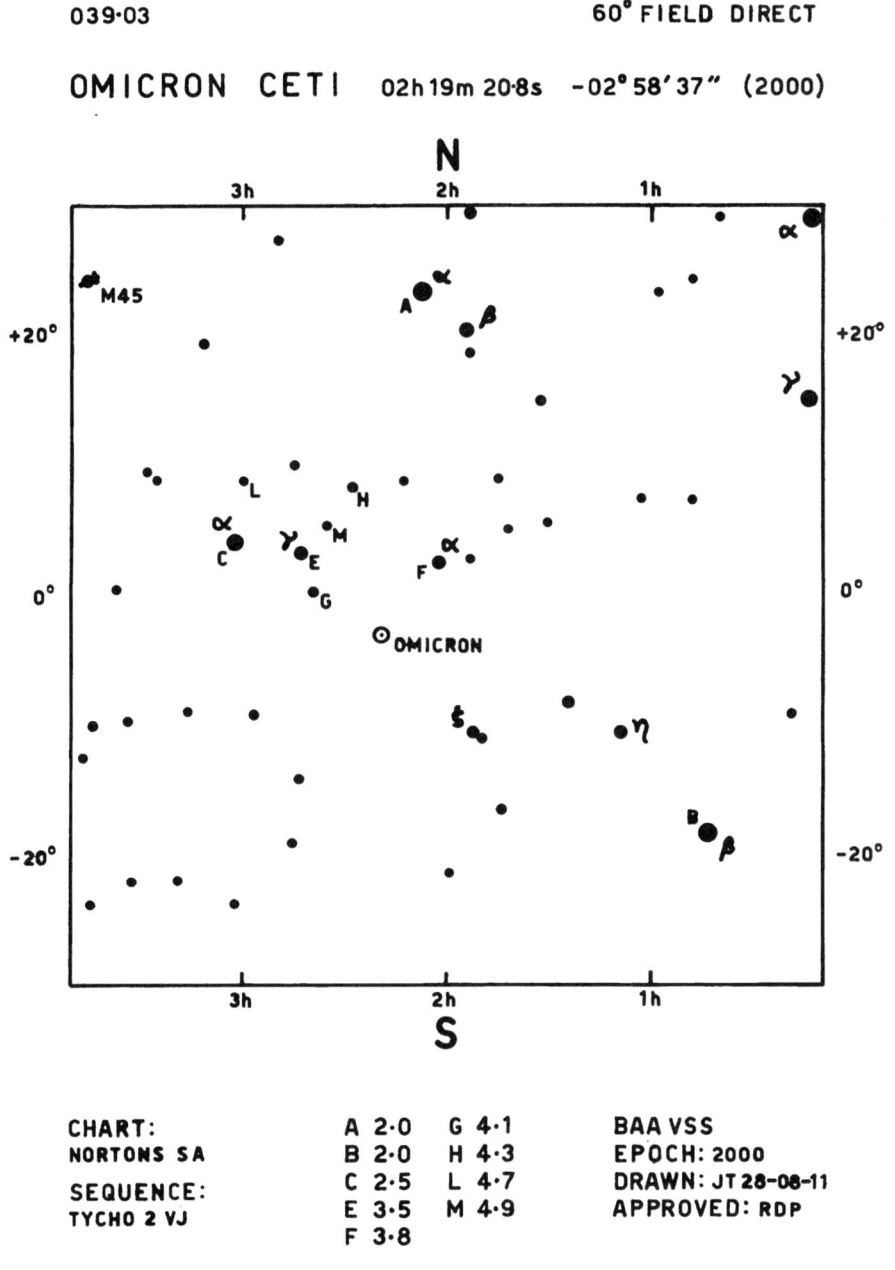

Fig. 14.5 Omicron Ceti

(Finder chart courtesy of the British Astronomical Association Variable Star Section)

R ANDROMEDAE
Type: M
Magnitude range: 5.8 to 15.2
Period: 409d
RA: 00.24.19   Dec: 38.34.37

Another beautiful long-period variable, and one of the first to be noticed after Mira, is R Andromedae. It has Mira-like features in that it is a red giant undergoing pulsations that make it a long-period variable star, but its spectrum reveals telltale signs of molecules such as zirconium monoxide, indicating that it is cooler than Mira and possibly a little older. Additional elements such as Technetium have been observed in the absorption lines of the spectra, revealing that underlying nucleosynthesis is ongoing in the outer envelopes of such distended red giants.

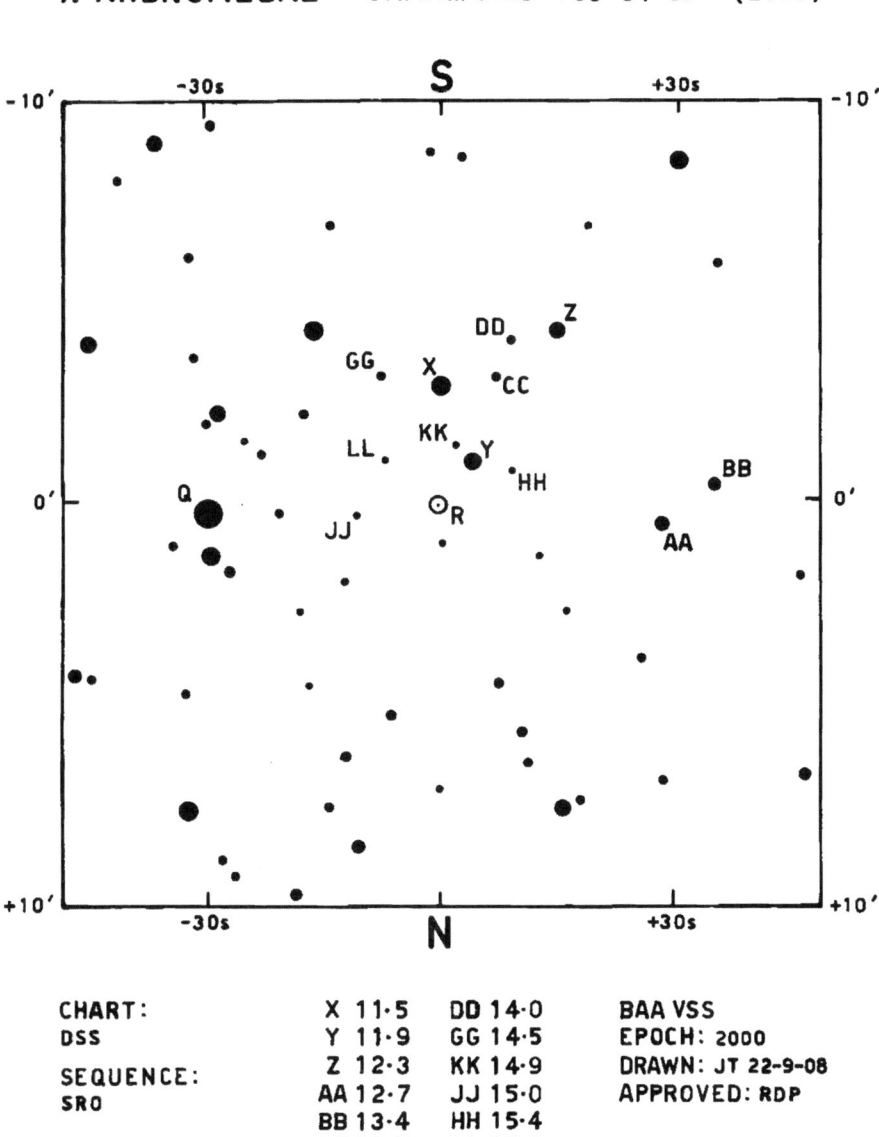

053·02   20' FIELD INVERTED

R ANDROMEDAE   00h 24m 01·9s +38° 34' 37" (2000)

Fig. 14.6 R Andromedae

(Finder chart courtesy of the British Astronomical Association Variable Star Section)

AG PEGASI
Type: NC
Magnitude range: 6 to 9.4
Period: –
RA: 21.51.01   Dec: 12.37.32

AG Pegasi is an unusual star in that it is symbiotic and undergoes regular bright outbursts that are termed "slow novae." Typically shining at 9th magnitude, the star can erupt without warning and display shell-like spectra reminiscent of the large star P Cygni.

Proof of its unusual activity stems from studies made over the 19th and 20th centuries. The spectrum of the hotter star, possibly a white dwarf star has changed drastically over 160 years, leading astronomers to surmise that its hotter component, originally a white dwarf, accumulated enough material from the donor giant star to begin burning hydrogen and enlarge and brighten into an A-type white supergiant around 1850. AG Pegasi demonstrated this spectrum in 1900 and had an estimated surface temperature of around 10,000 K. The system was measured to have a possible radius 16 times that of the Sun before the whole system changed over a long period to become a B-class star by 1920. It continued to show strange signs of change in its spectral output until by 1940, when it was classed as an O class object. It has continued to intrigue astronomers as its changes ended with the system being re-classed as a Wolf-Rayet star in 1975!

AG Pegasi has been described by some astronomers as the slowest nova ever recorded, but by the late 20th century, the hotter star has evolved into a hot subdwarf, and astronomers think that the whole system is returning to symbiotic status with an ordinary white dwarf.

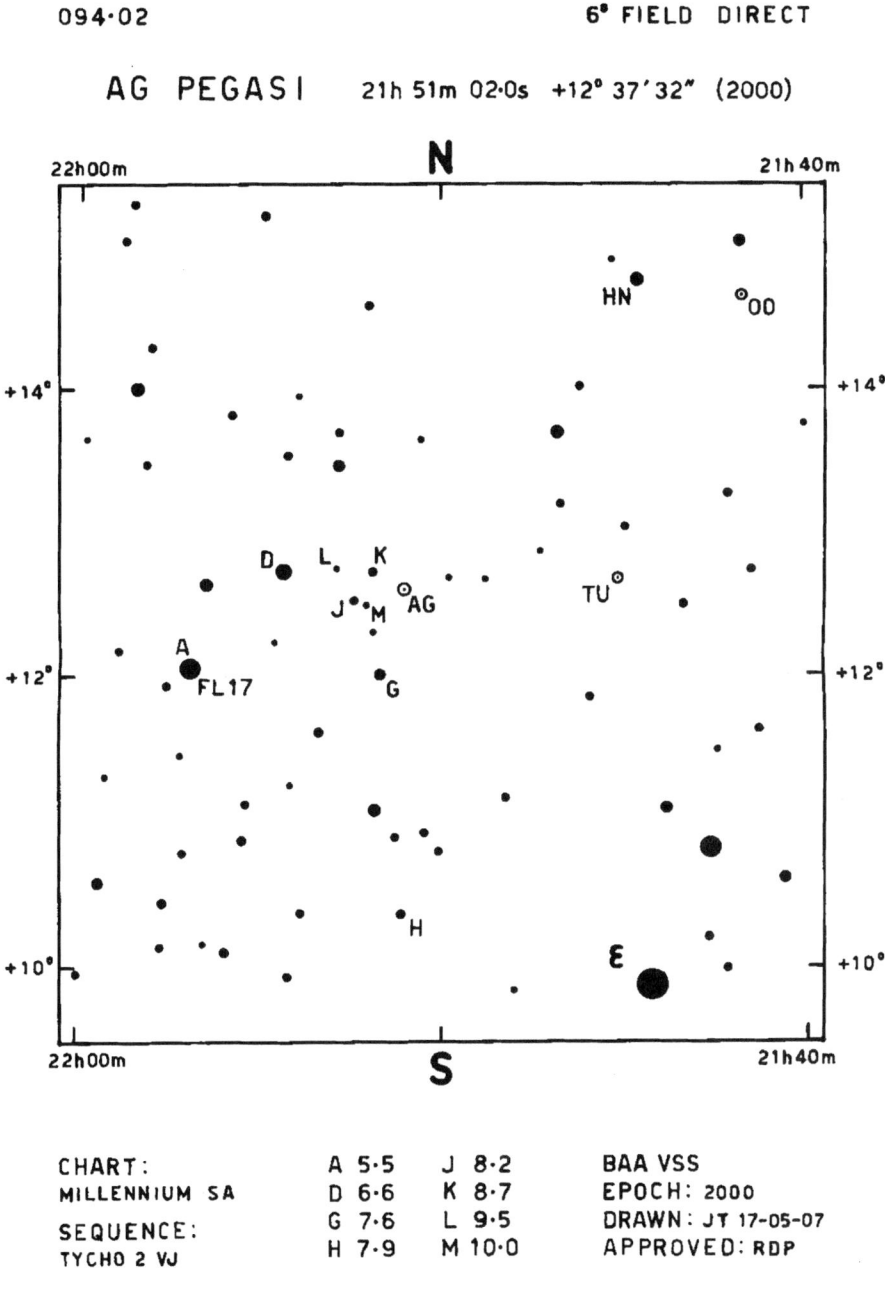

Fig. 14.7 AG Pegasi

(Finder chart courtesy of the British Astronomical Association Variable Star Section)

R AQUARII
Type: M
Magnitude range: 5.8 to 12.4
Period: 387d
RA: 23.43.49   Dec: −15.17.04

R Aquarii was the first variable star discovered in the constellation of the water carrier, and although initially classed as a Mira-type long-period variable, we now see that the star is a symbiotic type with a red giant primary and white dwarf companion star. It is one of the few stars that has had its radius measured, as its distance of 630 light years and its vast size make this possible for astronomers. The star is surrounded by a nebulous cloud of gas known as Cederblad 211. It is thought that this nebula is the remnant of a nova-like outburst that may have been observed by Japanese astronomers in the year A. D. 930. The nebula can be imaged by careful astronomers and is bright but small and dominated by its central star. The central region of the nebula show jets that are the result of material ejections that took place around 190 years ago. Better resolution of the nebula has also shown smaller structures indicating that the ejection is still ongoing and may be indicative of thermal pulsation or magnetic activity.

Given the broad range of magnitude R Aquarii displays, this star is an excellent one for beginners to watch out for.

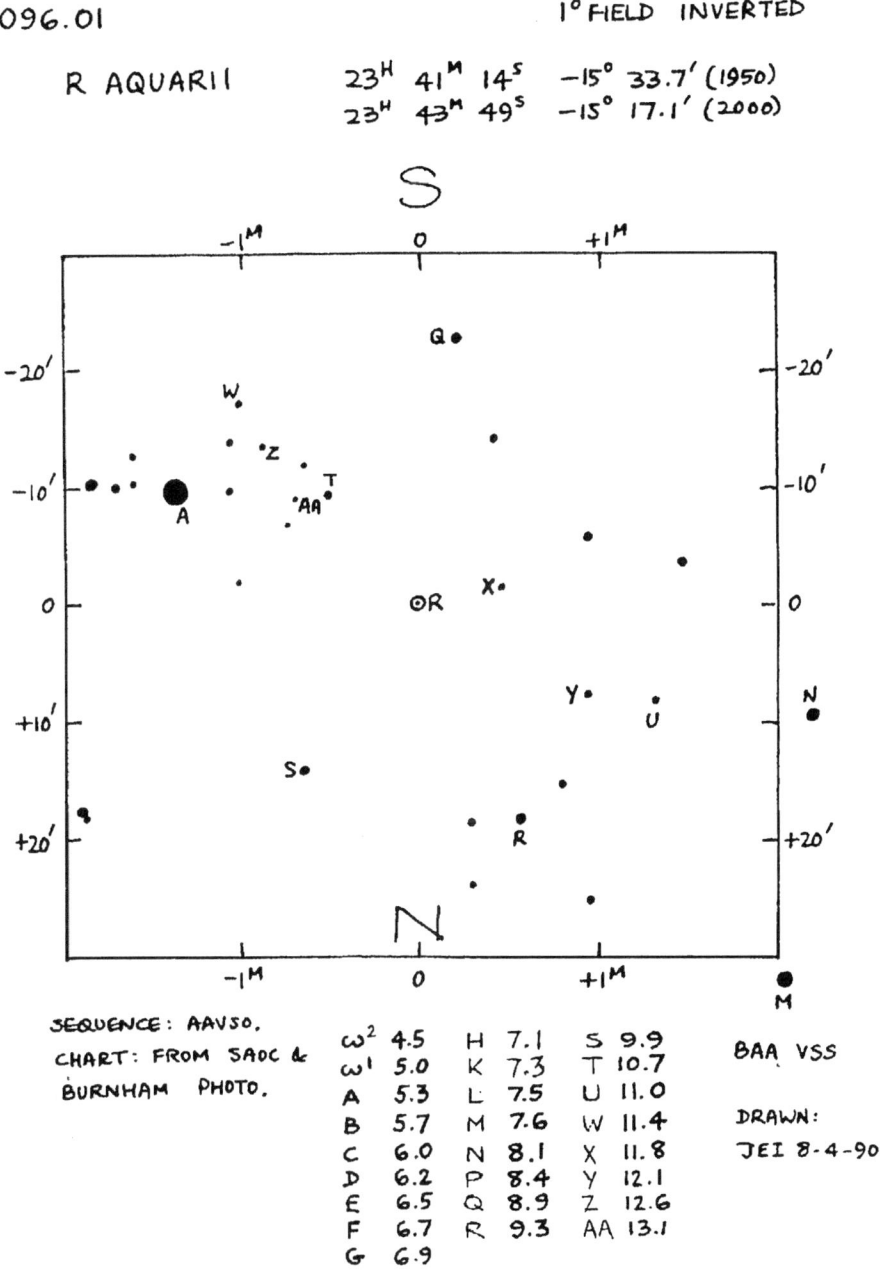

Fig. 14.8 R Aquarii

(Finder chart courtesy of the British Astronomical Association Variable Star Section)

For an additional list of stars that may prove of interest, please try the following telescopic variables. Finder charts for each are included here.

Star	Type	Range	Period	Frequency
Z And	ZAND	8.0–12.4p	NA	Nightly
X Cam	Mira	7.4–14.2	144d	5d-7d
RU Peg	UGSS	9.0–13.2	74d	Nightly
Y Lyn	SRc	6.5–8.4	110d	7d
Z Psc	SRb	7.0–7.9	144d	7d

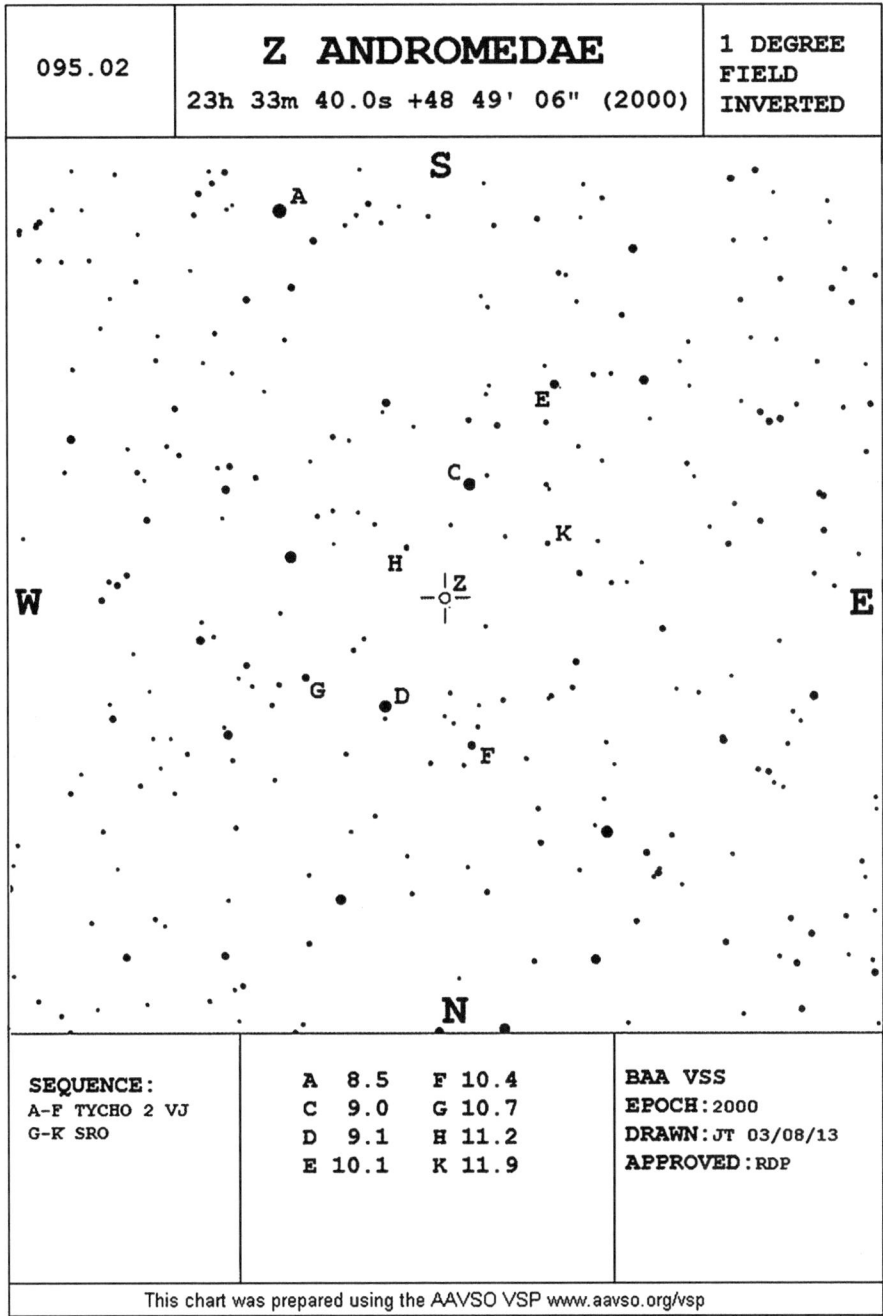

095.02	**Z ANDROMEDAE** 23h 33m 40.0s +48 49' 06" (2000)	1 DEGREE FIELD INVERTED

SEQUENCE: A-F TYCHO 2 VJ G-K SRO	A 8.5 F 10.4 C 9.0 G 10.7 D 9.1 H 11.2 E 10.1 K 11.9	BAA VSS EPOCH:2000 DRAWN:JT 03/08/13 APPROVED:RDP

This chart was prepared using the AAVSO VSP www.aavso.org/vsp

Fig. 14.9 Z Andromeda

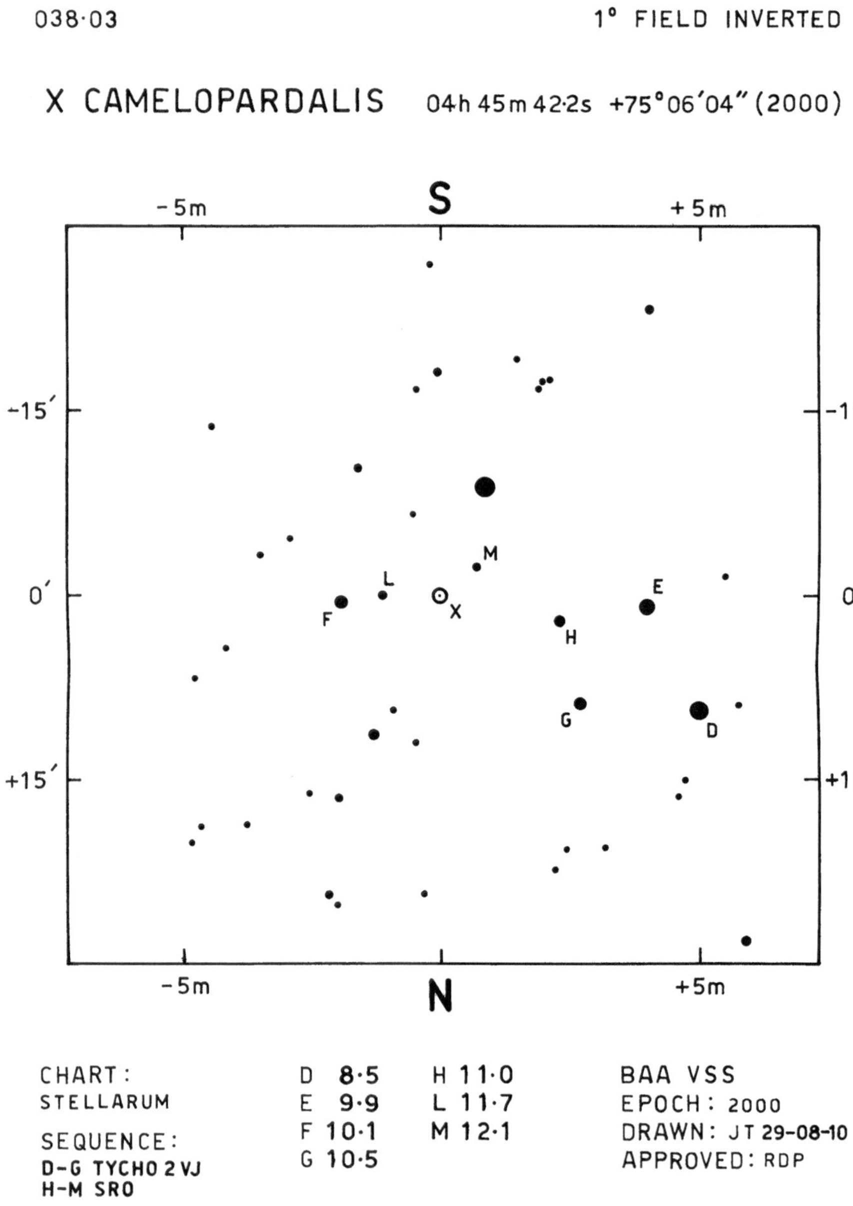
038·03
1° FIELD INVERTED
X CAMELOPARDALIS 04h 45m 42·2s +75°06′04″ (2000)
S
N
-5m
+5m
-15′
0′
+15′
X
D
E
F
G
H
L
M
CHART:
STELLARUM
SEQUENCE:
D-G TYCHO 2 VJ
H-M SRO
D 8·5
E 9·9
F 10·1
G 10·5
H 11·0
L 11·7
M 12·1
BAA VSS
EPOCH: 2000
DRAWN: JT 29-08-10
APPROVED: RDP

Fig. 14.10 X Camelopadralis

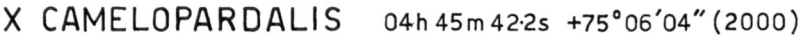

220912   RU Pegasi   1° FIELD

(1950) 22ʰ 11ᵐ·6 +12°27'

1°·8 p, 0°·5 N of Fl 31 Pegasi (4ᵐ·9)

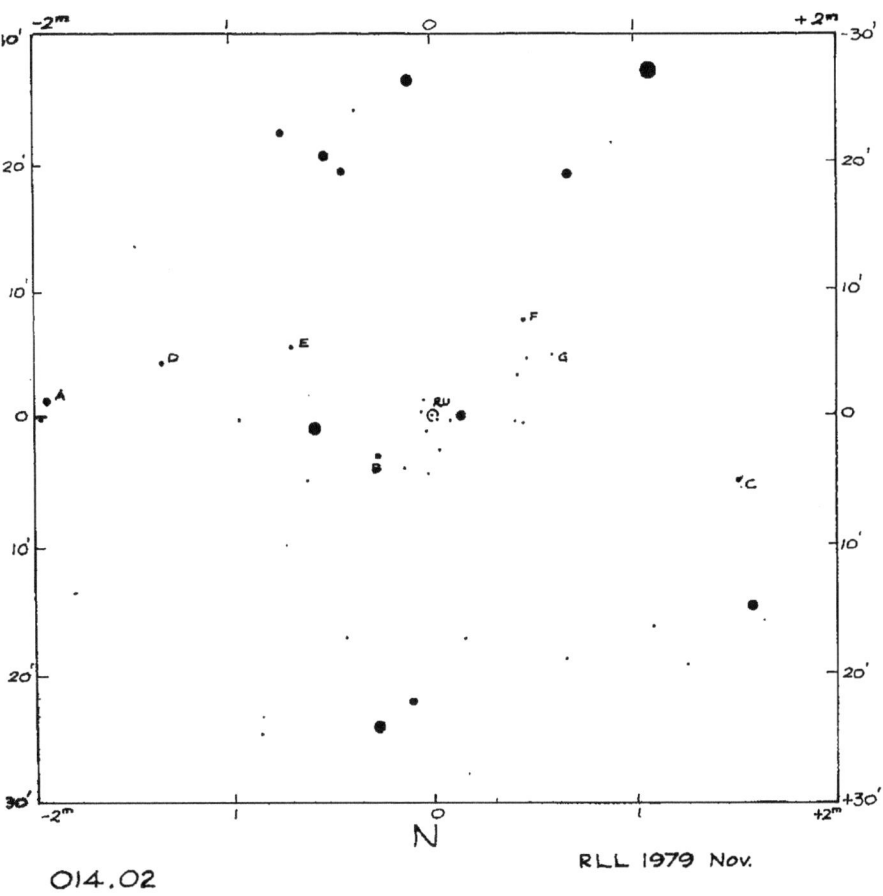

014.02

RLL 1979 Nov.

Fig. 14.11 RU Pegasi

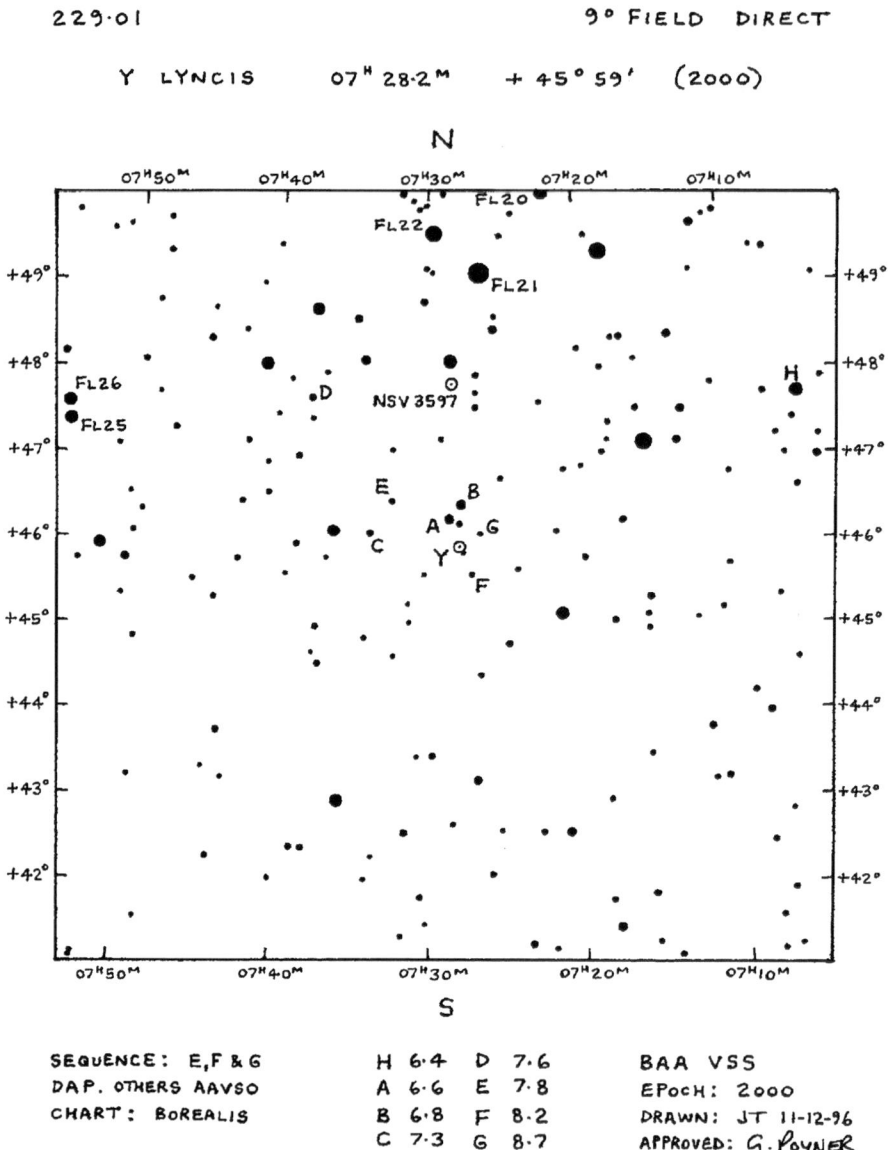

229.01   9° FIELD DIRECT

Y LYNCIS   07ᴴ 28.2ᴹ   + 45° 59' (2000)

SEQUENCE: E,F & G	H 6.4	D 7.6	BAA VSS
DAP. OTHERS AAVSO	A 6.6	E 7.8	EPOCH: 2000
CHART: BOREALIS	B 6.8	F 8.2	DRAWN: JT 11-12-96
	C 7.3	G 8.7	APPROVED: G. POYNER

Fig. 14.12 Y Lynxis

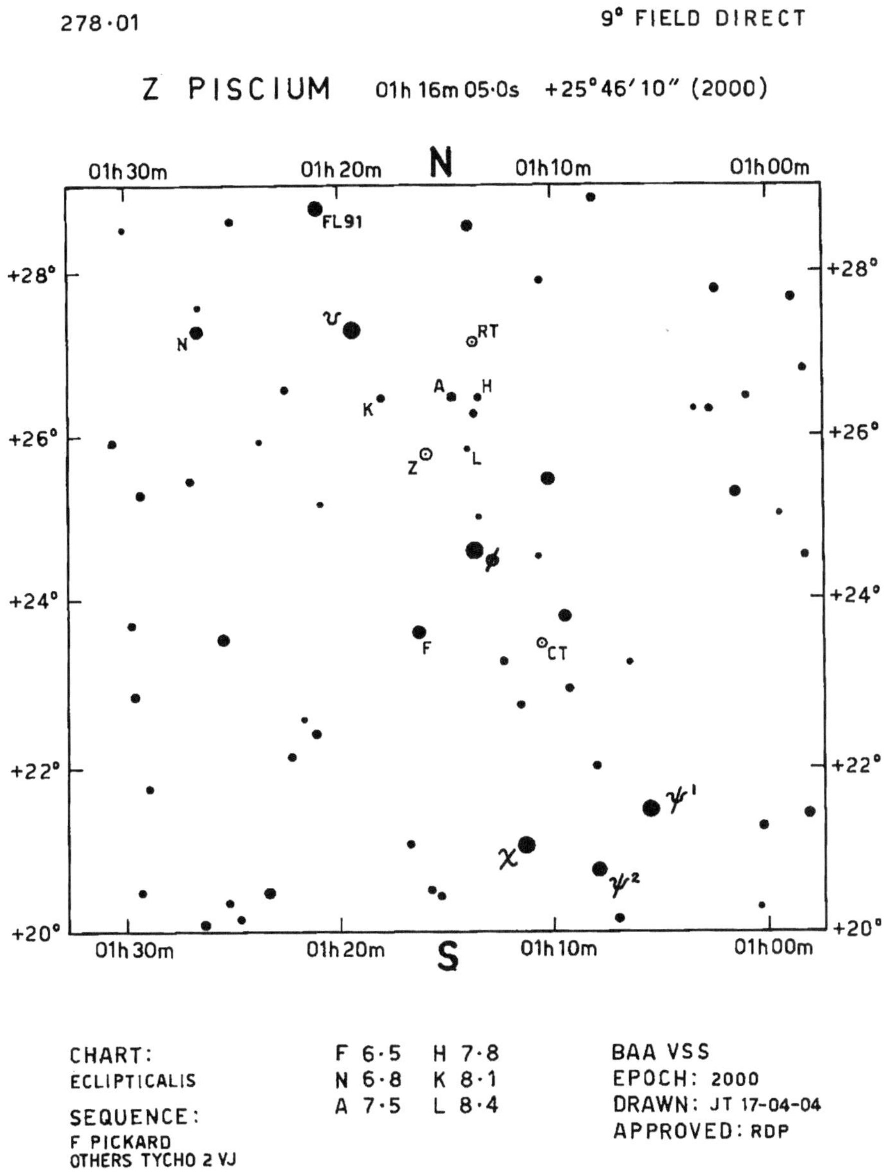
278·01
9° FIELD DIRECT
Z PISCIUM 01h 16m 05·0s +25°46′10″ (2000)
N
01h 30m
01h 20m
01h 10m
01h 00m
+28°
+26°
+24°
+22°
+20°
FL91
N
RT
A
H
K
Z
L
F
CT
S
CHART:
ECLIPTICALIS
SEQUENCE:
F PICKARD
OTHERS TYCHO 2 VJ
F 6·5 H 7·8
N 6·8 K 8·1
A 7·5 L 8·4
BAA VSS
EPOCH: 2000
DRAWN: JT 17-04-04
APPROVED: RDP

Fig. 14.13 Z Piscium

(Finder charts all courtesy of the British Astronomical Association Variable Star Section)

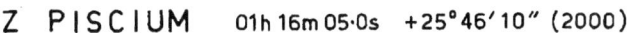

BU TAURI
Type: GCAS
Magnitude range: 4.8 to 5.5
Period: –
RA: 03.46.12   Dec: 23.59.1

An interesting star caught among the beautiful blue stars of the Pleiades, lying just above the bright star Pleione, this gamma Cassiopeia-type star is a slow variable in that its ejected shells may take several months if not years to make the star slowly change its light output. BU Tauri is also thought to have a close binary companion that can only be seen spectroscopically, but what role that star plays in its variability is unknown. This star, and Gamma Cassiopeia itself, was studied by Dr. John Griffiths in the UK, who showed that such stars have rapid rotation and the shell ejections may be a loss mechanism for Be-type stars as they probably have large equatorial bulges where the radiative output of some stars, married to centripetal forces, may eject matter. BU Tauri is also a source of faint X-ray emission.

034323   BU Tauri   4·8 – 5·5   ϒC   B8
(1950) 03ʰ 46ᵐ2 +23° 59′

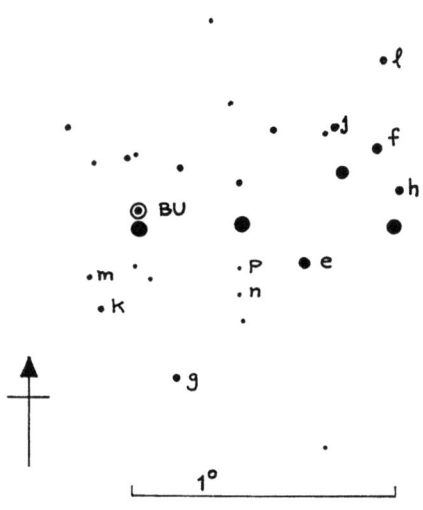

Fig. 14.14 Bu Tauri

Source:
BAA Handbook 1962

e 4·18   ℓ 5·65   n 6·99   BAA VSS
f 4·31   j 5·76   p 7·26
g 5·45   k 6·17
h 5·46   m 6·74

Revised JEI
1969 Aug 13
Redrawn MDT
1972 May 27 ϰ
1983 Oct 3

(Finder chart courtesy of the British Astronomical Association Variable Star Section)

SU TAURI
Type: RCB
Magnitude range: 9.1 to 17
Period: –
RA: 05.43.02   Dec: 19.02.04

SU Tauri is an R Coronae Borealis star that remains at a peak brightness until it drops dramatically over a period of hours or days to remain at minimum for an extended time. Like the prototype of the class, SU Tauri may remain constant between 9th and 10th magnitude for several years and then exhibit wild and unpredictable drops in brightness to 17th magnitude or fainter. In 2011 the AAVSO observer John Bortle noted that SU Tauri began another fade out after not quite recovering from its fade of the last few seasons. At present the star is slowly climbing back to maximum brightness but is still well below its usual magnitude level.

017·03   10′ FIELD INVERTED

SU TAURI   05h 49m 03·7s   +19° 04′ 22″ (2000)

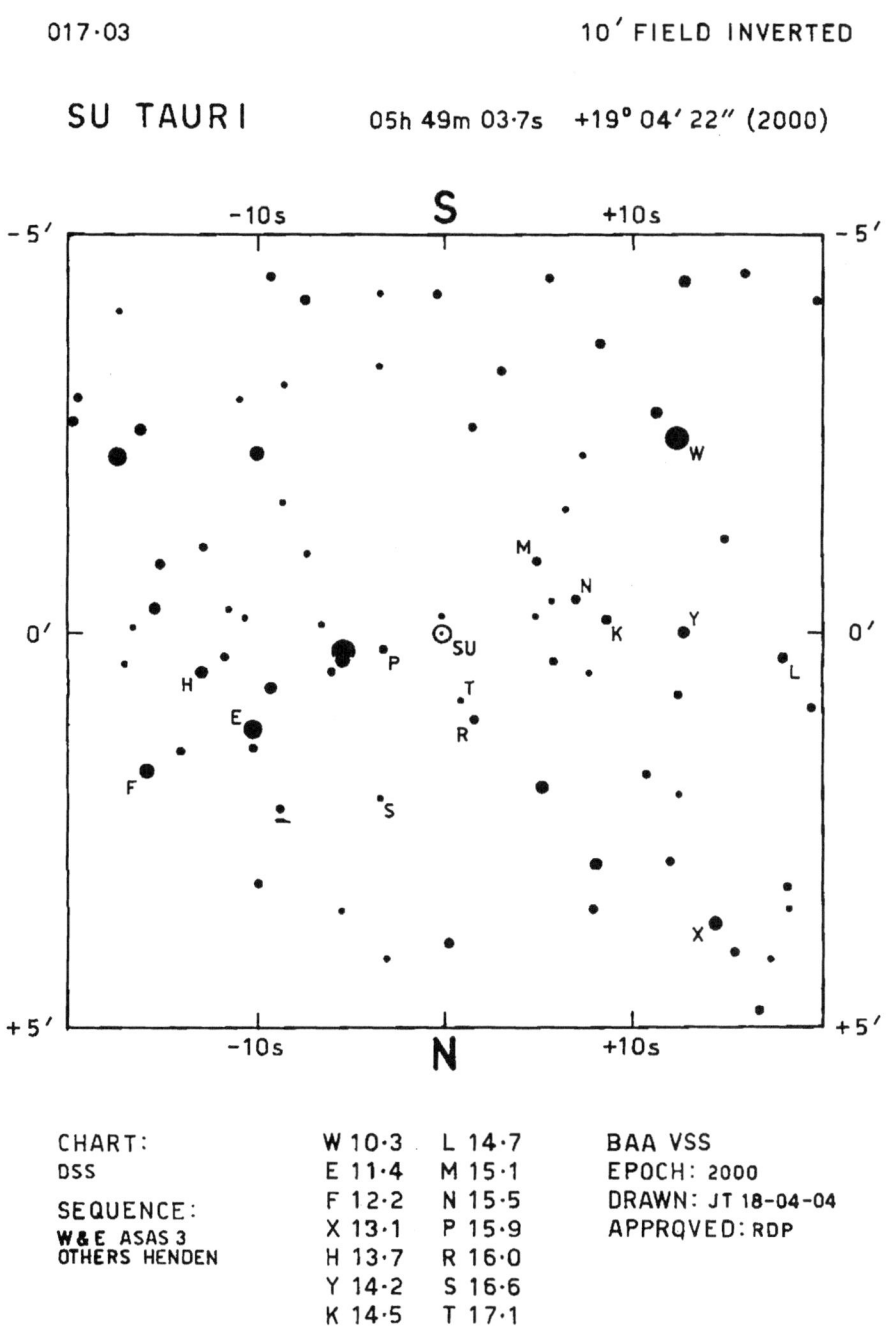

CHART:	W 10·3	L 14·7	BAA VSS
DSS	E 11·4	M 15·1	EPOCH: 2000
SEQUENCE:	F 12·2	N 15·5	DRAWN: JT 18-04-04
W&E ASAS 3	X 13·1	P 15·9	APPRQVED: RDP
OTHERS HENDEN	H 13·7	R 16·0	
	Y 14·2	S 16·6	
	K 14·5	T 17·1	

Fig. 14.15 Su Tauri

(Finder chart courtesy of the British Astronomical Association Variable Star Section)

T TAURI
Type: INT
Magnitude range: 9.3 to 13.5
Period: –
RA: 04.21.59   Dec: 19.32.06

The prototype of a class of variable young sun-like stars, T Tauri is associated with Hind's variable nebula NGC 1555, which lies very close by. The star is less than one million years old and is in fact three stars in a triple system. Studies by the Very Large Array several years ago have found that T Tauri itself encountered one of these companions, and the interaction has flung it out of the system as a runaway star. T Tauri is about 430 light years away, and although it lies among the Hyades stars is considerably more distant than this close cluster.

The star itself is quite faint, mostly, and shows up well in photographs of NGC 1555. It has flare ups that make its brightness increase, but these are irregular, and the maxima are different with every outburst. The brightest it has been recorded is magnitude 8.9, so a good aperture telescope is needed to keep an eye on it.

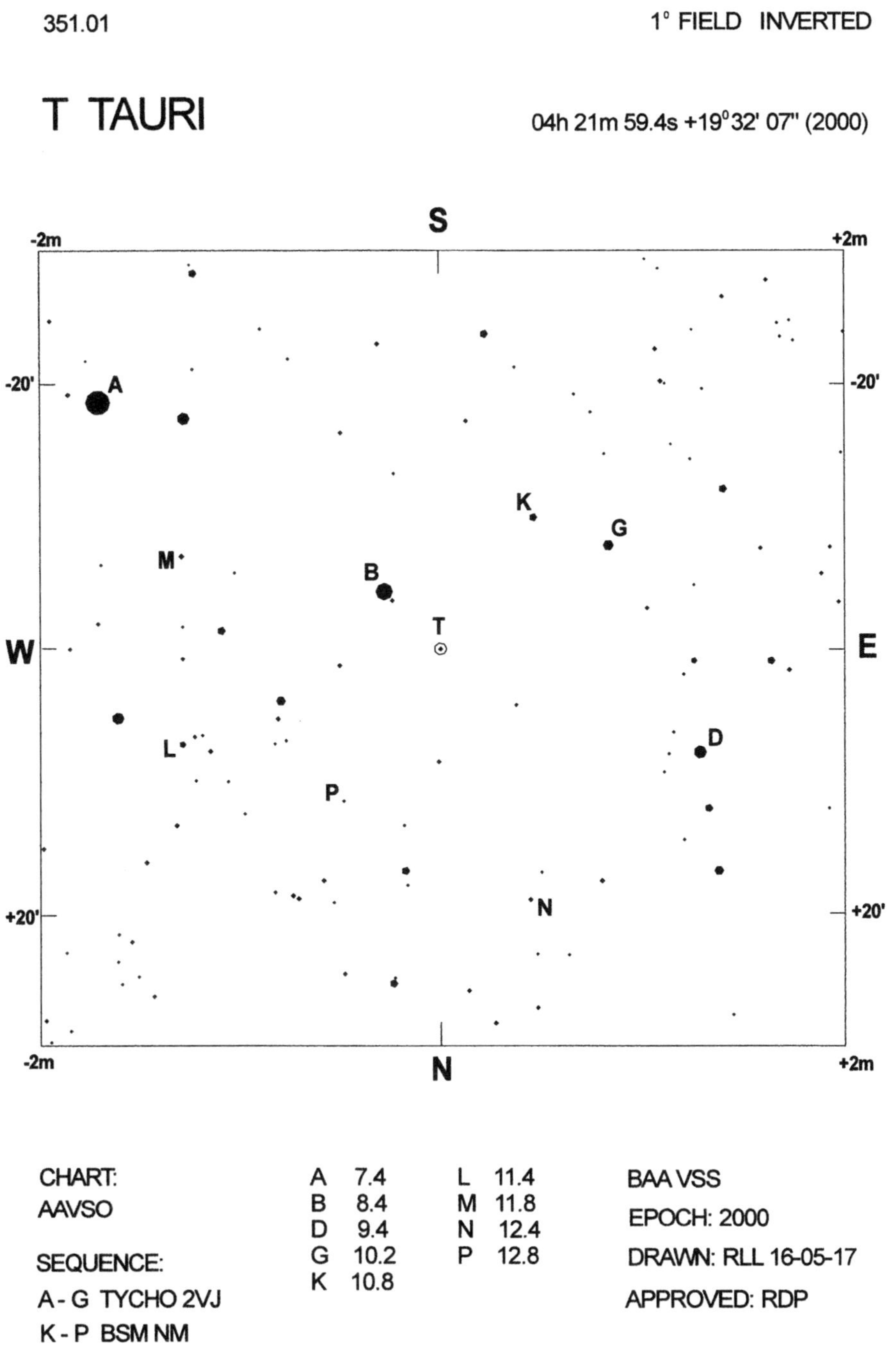

351.01   1° FIELD   INVERTED

T TAURI

04h 21m 59.4s +19°32' 07" (2000)

CHART:	A	7.4	L	11.4	BAA VSS
AAVSO	B	8.4	M	11.8	EPOCH: 2000
	D	9.4	N	12.4	
SEQUENCE:	G	10.2	P	12.8	DRAWN: RLL 16-05-17
	K	10.8			
A - G TYCHO 2VJ					APPROVED: RDP
K - P BSM NM					

Fig. 14.16 T Tauri

(Finder chart courtesy of the British Astronomical Association Variable Star Section)

S PERSEI
Type: SRC
Magnitude range: 7.9 to 12
Period: 822d
RA: 02.22.51   Dec: 58.31.11

One of the largest supergiants known, S Persei can be found close to the double cluster of stars in the sword handle of Perseus. Part of the Perseus OB1 association, this red supergiant is an evolved star that is extremely luminous and has a diameter over 1,200 times that of the Sun. The star varies slowly and irregularly in maximum and minima over a period of two years, but since it is a semi-regular star, recordings of its light curve are essential to monitoring this object. At its brightest it can be seen in binoculars but requires a good telescope at minimum. It is also a binary star, with its companion a good way from it. The companion is an A-type 11th magnitude star about 69 arcseconds away.

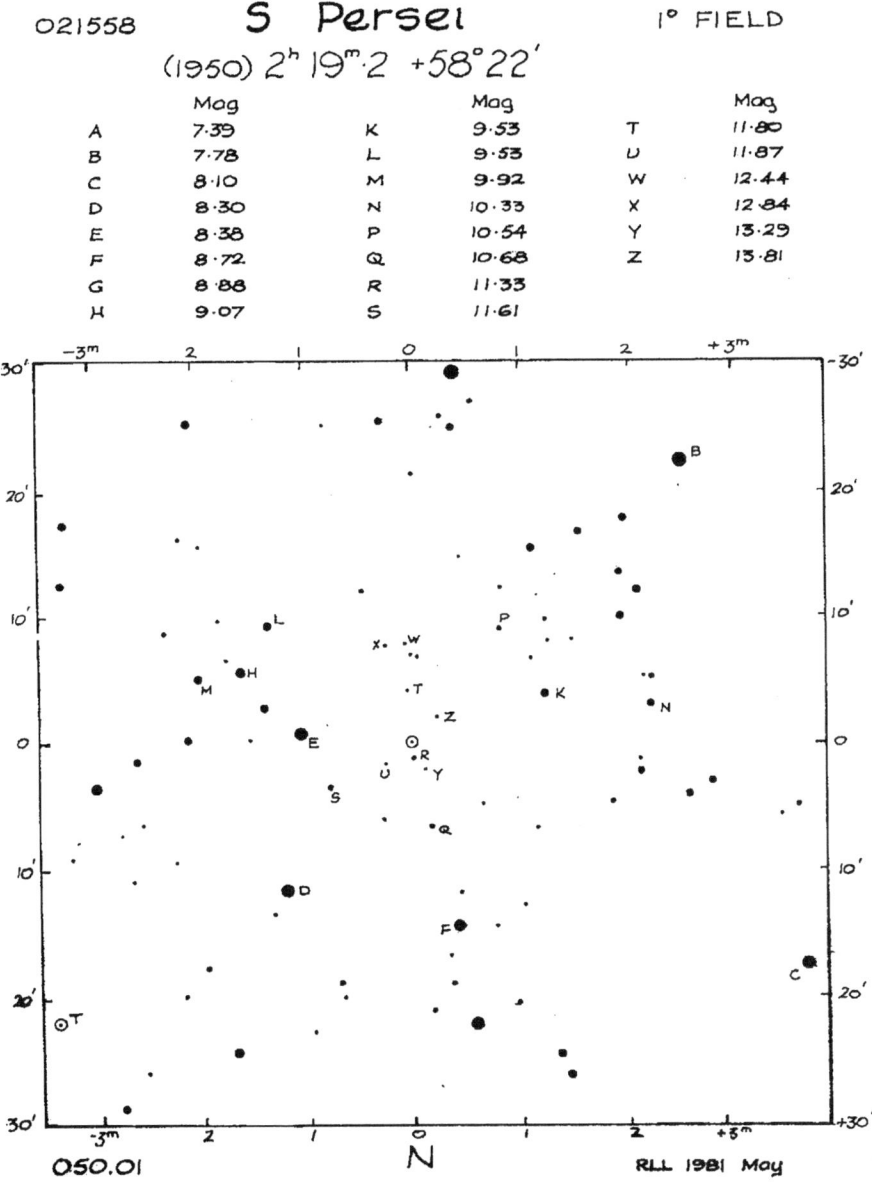

Fig. 14.17 S Persei

(Finder chart courtesy of the British Astronomical Association Variable Star Section)

X PERSEI
Type: Be/X
Magnitude range: 6 to 7
Period: –
RA: 03.55.23   Dec: 31.02.45

X Persei is one of the X-ray binary class as the large B-type primary star is in orbit with a neutron star companion. The magnitude range of this variable is not large, but the proximity of such a dense object makes this an interesting system. In 1989 and again in 1990, the spectrum of X Persei changed from a Be star to a normal B-class star while it faded significantly, probably due to the loss of the accretion disc. The disc has since reformed and shows strong emission lines in spectra taken in 2015 and 2017.

The neutron star companion is a radio pulsar and X-ray source in the Uhuru catalogue. The pulsar has shown period changes that are associated with mass transfer from the more massive primary star, though it is unclear whether an accretion disc has formed or if the mass transfer is direct. Between 1973 and 1979, the pulsar was seen to increase its rate of spin, associated with a strong X-ray flare and presumed strong mass transfer. Since then the spin has been slowing despite small X-ray flares, so there may be direct transfer onto the pole of the neutron star. Given its extreme brightness, the variability can be followed in binoculars, too.

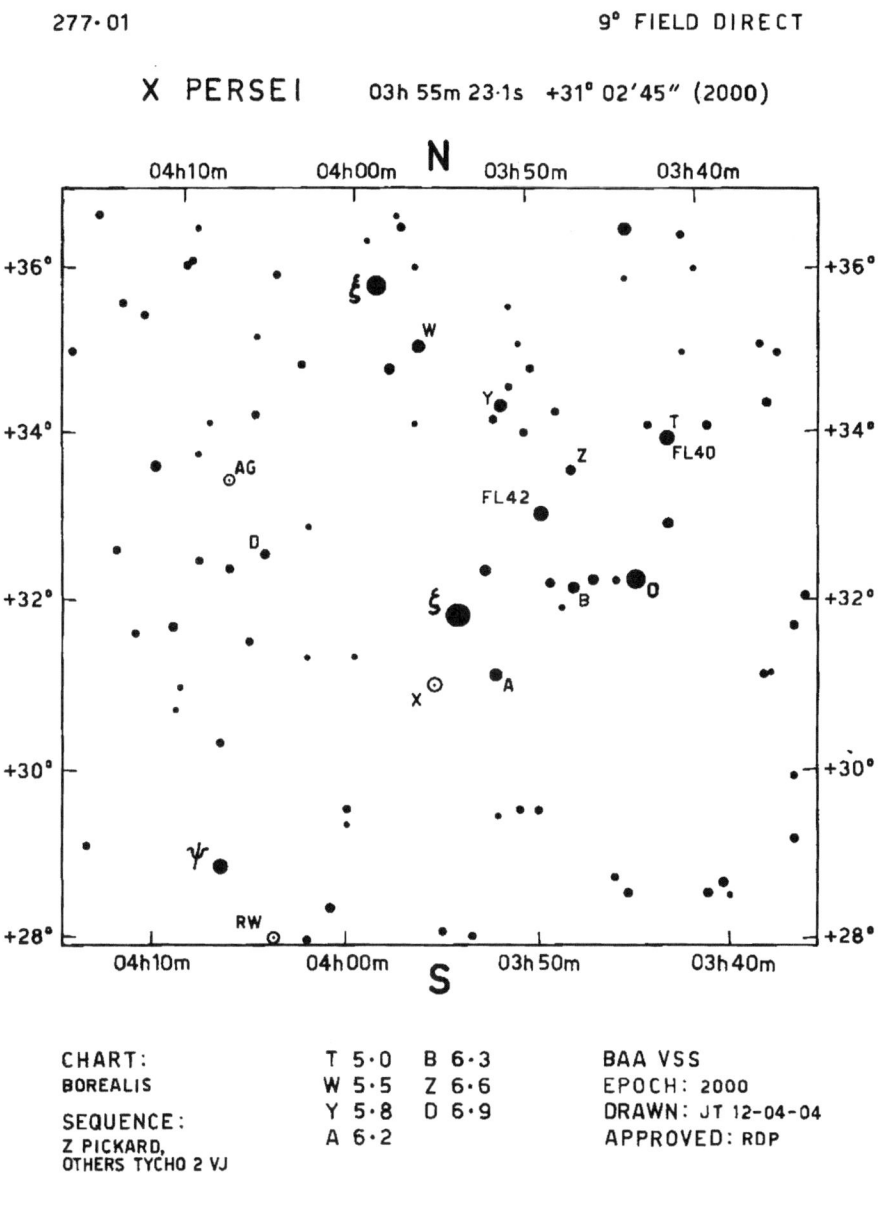

Fig. 14.18 X Persei

(Finder chart courtesy of the British Astronomical Association Variable Star Section)

U ORIONIS
Type: M
Magnitude range: 4.8 to 13
Period: 368d
RA: 05.55.49   Dec: 20.10.30

A beautiful Mira-type long-period variable, U Orionis has been observed for over 130 years, discovered in 1885 by the Irish astronomer, J. E. Gore. It is famous in that it is the first variable star to be identified by means of its spectra. U Orionis is 7,000 times more luminous than our Sun and lies almost 1,000 light years away. It is a very late M-class star of type M8 and has a surface temperature of only 2,700 K. Seen in a small refractor the star is easily identified, as it is quite red. It has a large magnitude range that brings it to naked-eye visibility and can be followed quite well across its entire magnitude range if you have access to a large telescope.

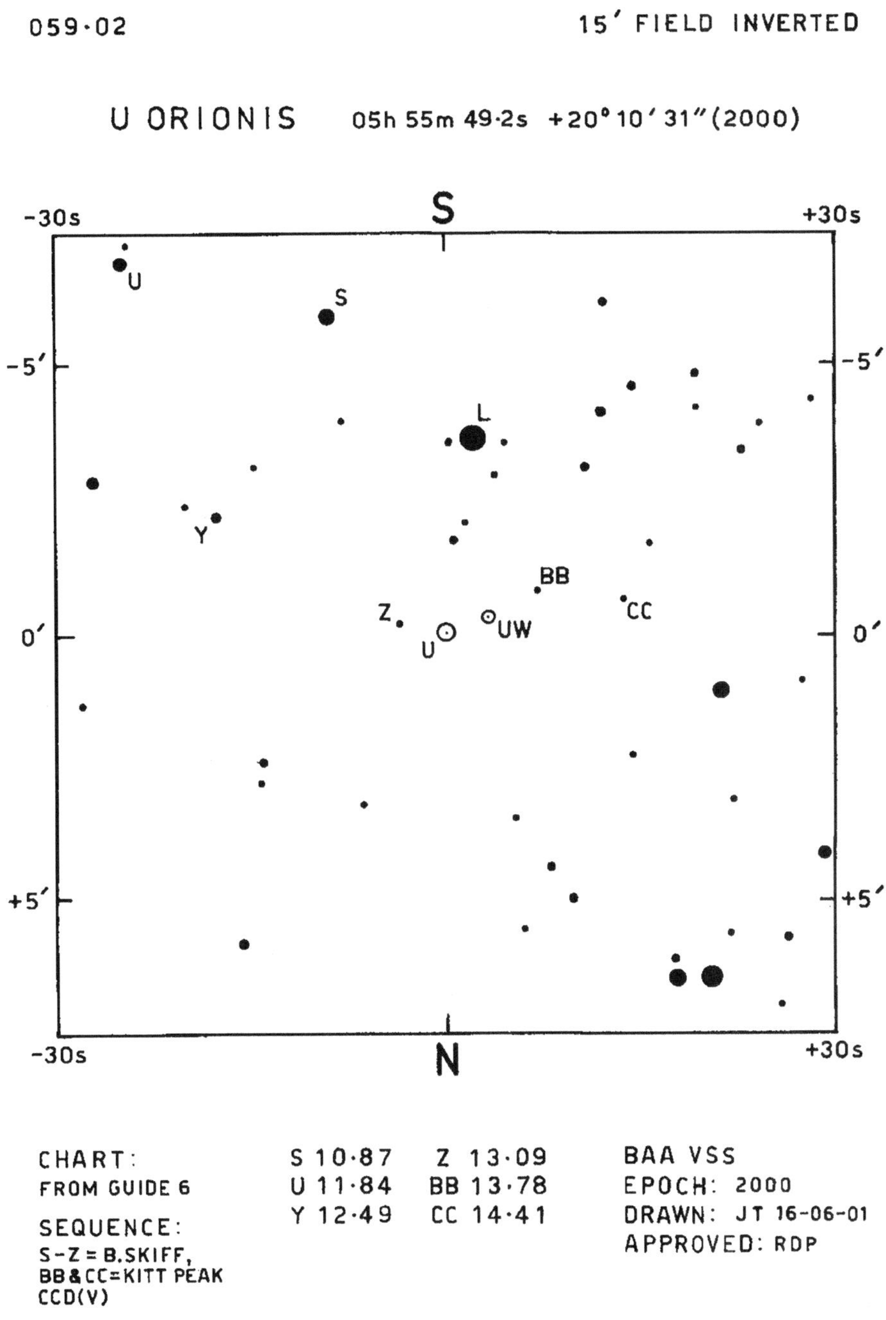

059·02
15′ FIELD INVERTED
U ORIONIS 05h 55m 49·2s +20° 10′ 31″(2000)
S
N
-30s
+30s
-5′
0′
+5′
U
S
L
Y
BB
CC
Z
U
UW
CHART:
FROM GUIDE 6
SEQUENCE:
S-Z = B.SKIFF,
BB & CC=KITT PEAK
CCD(V)
S 10·87 Z 13·09
U 11·84 BB 13·78
Y 12·49 CC 14·41
BAA VSS
EPOCH: 2000
DRAWN: JT 16-06-01
APPROVED: RDP

Fig. 14.19 U Orionis

(Finder chart courtesy of the British Astronomical Association Variable Star Section)

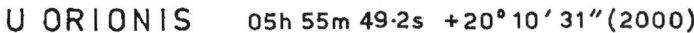

BX MONOCEROTIS
Type: Unique
Magnitude range: 9.3 to 13.4
Period: 1374d
RA: 07.22.53   Dec: –03.29.8

BX Monocerotis was originally classed as a long-period variable of the Mira type but is now recognized as a symbiotic star that has a slow rise to brightness and fading due to the orbital mechanics of the system. Its 1,374 day period is actually the period of rotation of the companion star in the system, and the star may be an eclipsing binary, too. There are small scale fluctuations in brightness that may be captured photometrically, which appear to be due to gas exchange between the primary red giant and the white dwarf companion, creating a small hotspot that is possibly responsible for the brief peaks in light output.

Again, this is a faint star for most amateurs but could be monitored as a challenge by more experienced astronomers.

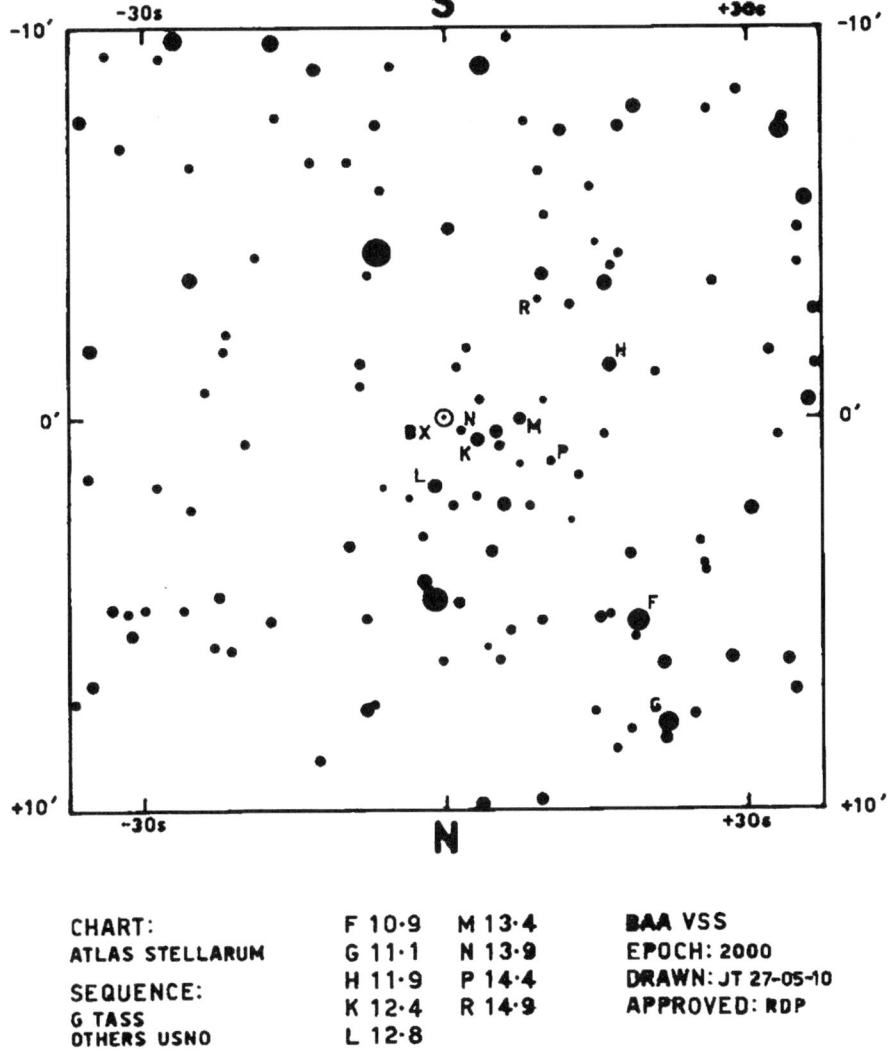

Fig. 14.20 BX Monocerotis

(Finder chart courtesy of the British Astronomical Association Variable Star Section)

U MONOCEROTIS
Type: RVB
Magnitude range: 6.1 to 8.8
Period: 91.3d
RA: 07 30.47   Dec: −09.46.36

An RV Tauri-type variable, U Monocerotis is a yellow-white giant star that can often be seen with the naked eye lying between Sirius and Procyon in the winter Milky Way, but drops below naked-eye visibility at deep minima. It lies only two degrees west of alpha Monocerotis, which at fourth magnitude is the brightest star in the dismal constellation of the unicorn. At its brightest, U Mon can reach magnitude 5.45. At a shallow minimum it drops to about magnitude 6.0, but at its deepest minima it is below magnitude 7.5. The period is given as 92.2 days, although this varies slightly from cycle to cycle. The brightness of the main pulsations varies over a long secondary period.

U Monocerotis is an interesting binary system with a dusty ring surrounding both stars. However, no one has ever seen the companion star, though its presence is inferred by radial velocity changes, positing that it orbits every 2,597 days. The stars could be occulted by a circumstellar disc that surrounds them both. As the magnitude range is not huge, the star may be followed in binoculars.

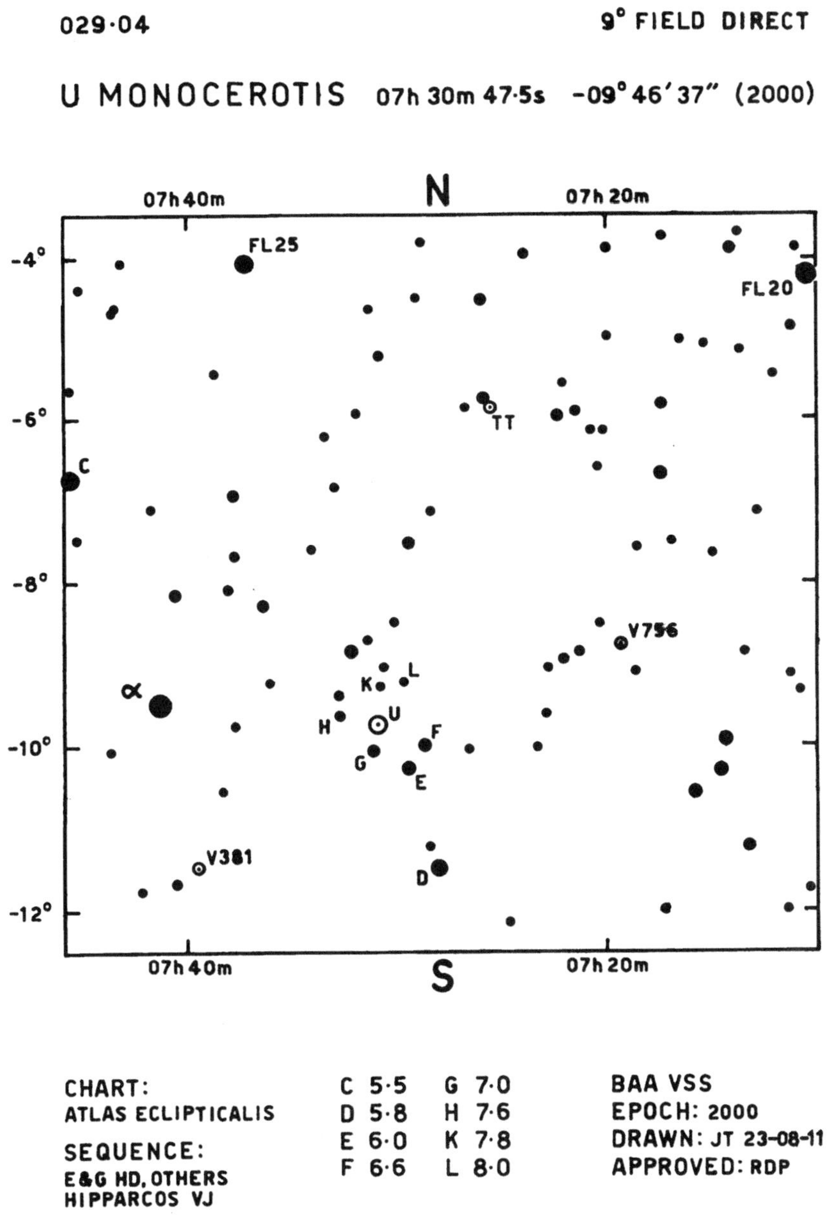

Fig. 14.21 U Monocerotis

(Finder chart courtesy of the British Astronomical Association Variable Star Section)

EPSILON AURIGAE
Type: Unique
Magnitude range: 2.9 to 3.8
Period: 9896d
RA: 05.51.08   Dec: 43.49.23

Epsilon Aurigae is one of the most fascinating variable stars in the sky, but you will have to wait for the next eclipse! The system is an unusual eclipsing binary comprised of an F0 star and a strange companion, which is generally accepted to be a huge dark disc orbiting an unknown object, possibly a binary system of two small B-type stars. The distance to the system is still a subject of debate, but modern estimates place it approximately 2,000 light years away.

Epsilon Aurigae was first suspected to be a variable star when German astronomer Johann Heinrich Fritsch observed it in 1821. Later observations reinforced Fritsch's initial suspicions and attracted attention to the star. The astronomer Hans Ludendorff was the first to study it in great detail, and his work revealed that the system was an eclipsing binary variable, a view that has been reinforced by many observations since.

About every 27 years, Epsilon Aurigae's brightness drops from a visual magnitude of 2.92 to magnitude 3.83. Observations over the last two eclipses show that this dimming lasts for between 640 and 730 days. In addition to the eclipses, the system also has a low amplitude pulsation, with a period of around 66 days, though this requires further observations to prove.

The eclipsing companion has been subject to much debate, since the object does not emit as much light as is expected for an object its size. During the last eclipse, in 2008, the most popularly accepted model for this companion object was a binary star system surrounded by a massive, opaque disc of dust. Though unusual, it is the best fit for the observations, and theories speculating that the object is a large, semitransparent star or even a black hole have since been discarded. The next eclipse is in the years 2035–36.

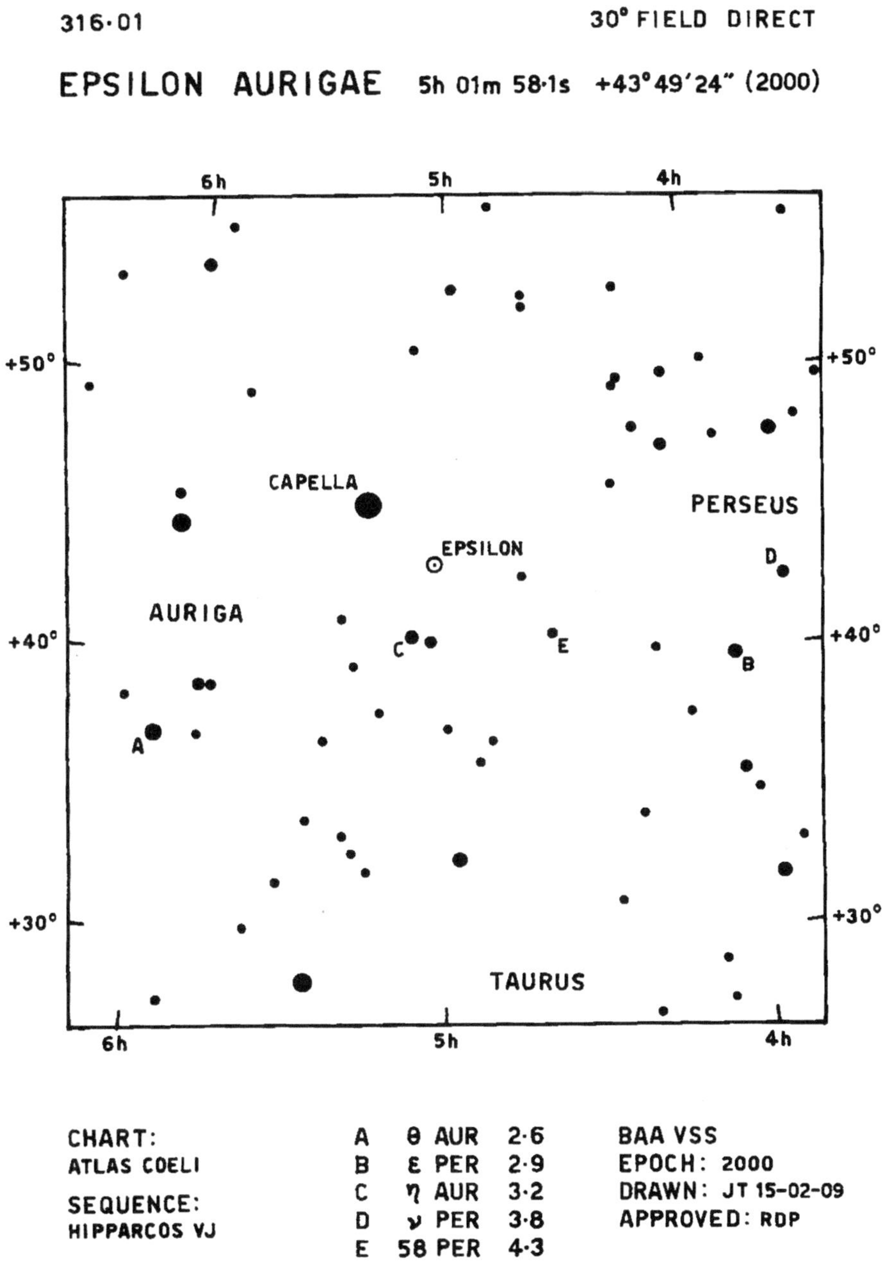

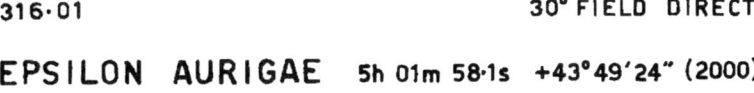

316·01
30° FIELD DIRECT
EPSILON AURIGAE 5h 01m 58·1s +43°49′24″ (2000)
6h
5h
4h
+50°
+40°
+30°
CAPELLA
EPSILON
PERSEUS
AURIGA
TAURUS
A
B
C
D
E
CHART:
ATLAS COELI
SEQUENCE:
HIPPARCOS VJ
A θ AUR 2·6
B ε PER 2·9
C η AUR 3·2
D ν PER 3·8
E 58 PER 4·3
BAA VSS
EPOCH: 2000
DRAWN: JT 15-02-09
APPROVED: RDP

Fig. 14.22 Epsilon Aurigae

(Finder chart courtesy of the British Astronomical Association Variable Star Section)

ETA GEMINORUM
Type: SRA/EA
Magnitude range: 3.2 to 3.9
Period: 233d
RA: 06.14.52   Dec: 22.30.24

This lovely triple star is also an eclipsing variable and a semi-regular variable, too. The variations in its output were first noticed by the astronomer Julius Schmidt in 1865. He described the star as having a long maxima of constant brightness and a minimum of greatly varying size and shape in its light curve, corresponding to a period of 231 days. The eclipse period has been set by modern observers at about eight years, similar to the orbit of an unseen companion. The eclipses themselves seem open to interpretation, as they have been questioned by astronomers many times, so at predicted eclipses, it is always worthwhile watching the star.

The semi-regular variations have been classified as type SRA, indicating relatively predictable periodicity with some variations in amplitude and light curve shape. Eta Geminorum is considered to be very similar to Mira-type variable stars but with smaller amplitudes. Many long-period variables show long secondary periods, typically ten times longer than the main period, but these changes have not been detected for η Geminorum. The main period of variability has been refined to an average of 234 days. Eclipse records and timings are available from the AAVSO. Although the light amplitude is not great in magnitude, nevertheless Eta Geminorum is quite an easy star to watch and gain experience from.

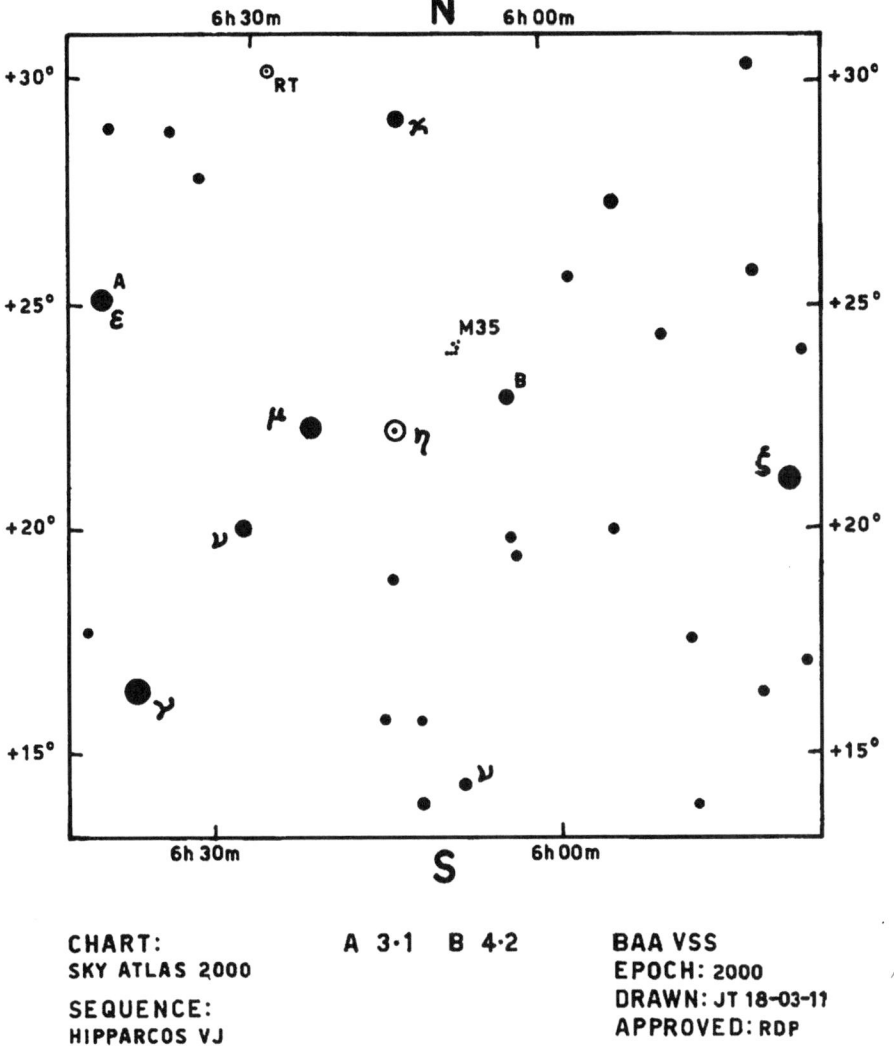

Fig. 14.23 Eta Geminorum

(Finder chart courtesy of the British Astronomical Association Variable Star Section)

U GEMINORUM
Type: UGSS
Magnitude range: 8.2 to 14.9
Period: 105d
RA: 07.55.05   Dec: 22.00.05

U Geminorum is the prototype of a class of dwarf novae with a white dwarf and a red dwarf in close proximity to each other. Discovered by J. R. Hind in 1855 as a novae, the star faded quickly but was seen again shortly after in outburst. The U Geminorum system has a very short orbital period of 4 hours and 11 minutes, and it is this orbit alone that makes the system an eclipsing variable, as each star transits with each orbit. Though the average interval between outbursts is 102 days, the period of variability is in fact highly irregular, varying from as little as 62 days to as long as 257. It is unknown when U Geminorum will go into outburst, so the star requires careful observation. As with many dwarf stars, the distance is not well known and may be as close as 170 light years or as distant as 300 light years.

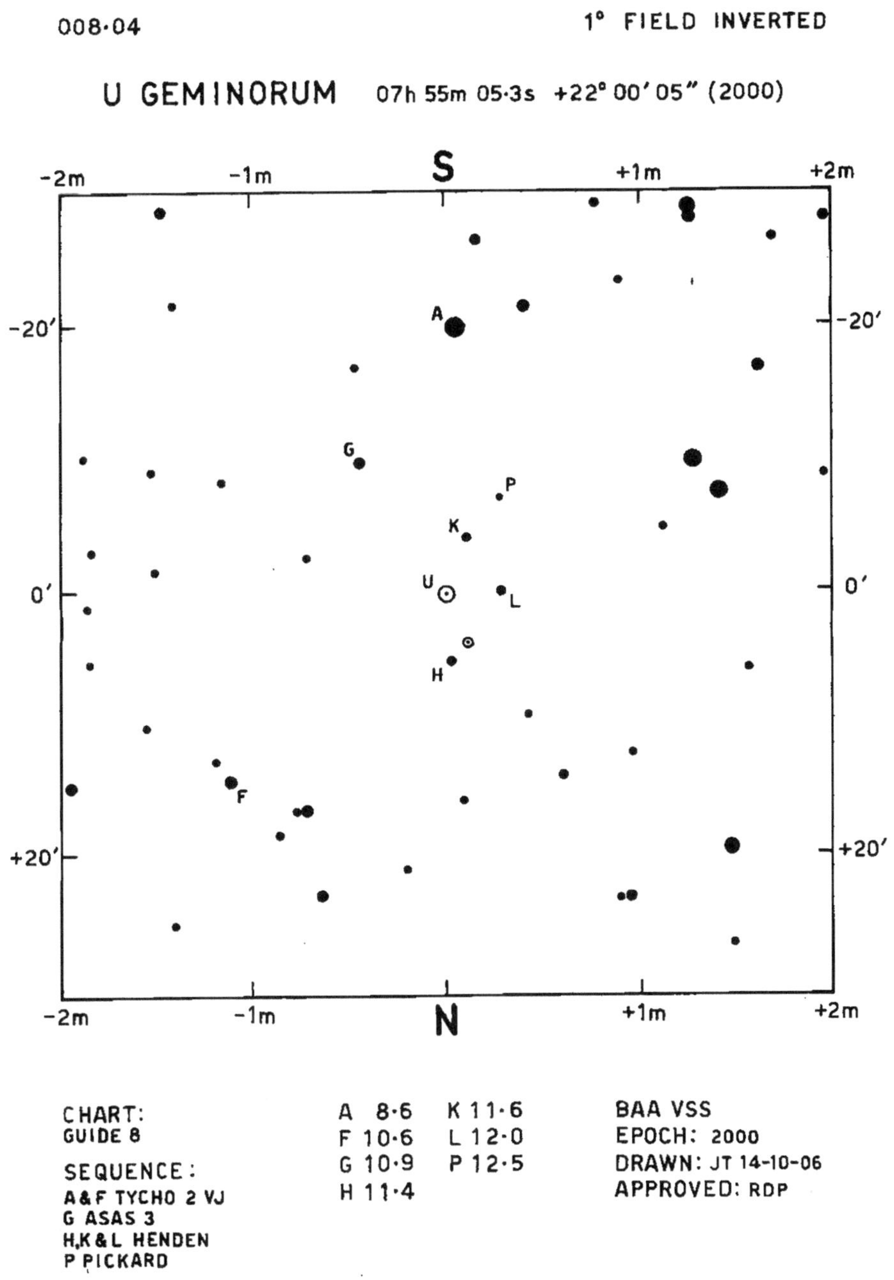

008·04   1° FIELD INVERTED

U GEMINORUM   07h 55m 05·3s +22° 00′ 05″ (2000)

CHART:	A 8·6	K 11·6	BAA VSS
GUIDE 8	F 10·6	L 12·0	EPOCH: 2000
SEQUENCE:	G 10·9	P 12·5	DRAWN: JT 14-10-06
A&F TYCHO 2 VJ	H 11·4		APPROVED: RDP
G ASAS 3			
H,K&L HENDEN			
P PICKARD			

Fig. 14.24 U Geminorum

(Finder chart courtesy of the British Astronomical Association Variable Star Section)

For an additional list of stars that may prove of interest, please try the following telescopic and binocular variables. Finder charts for each are included here.

Star	Type	Range	Period	Frequency
AX Per	ZAND+E	8.0–13.0	NA	Nightly
RV Tau	RVB	8.8–11.0	77d	5d
BU Gem	Lc	5.7–8.1	47d?	5d
X Cnc	SRb	5.6–7.5	195d	7d

073·02   20′ FIELD INVERTED

AX PERSEI   01h 36m 22·7s   +54° 15′ 02″ (2000)

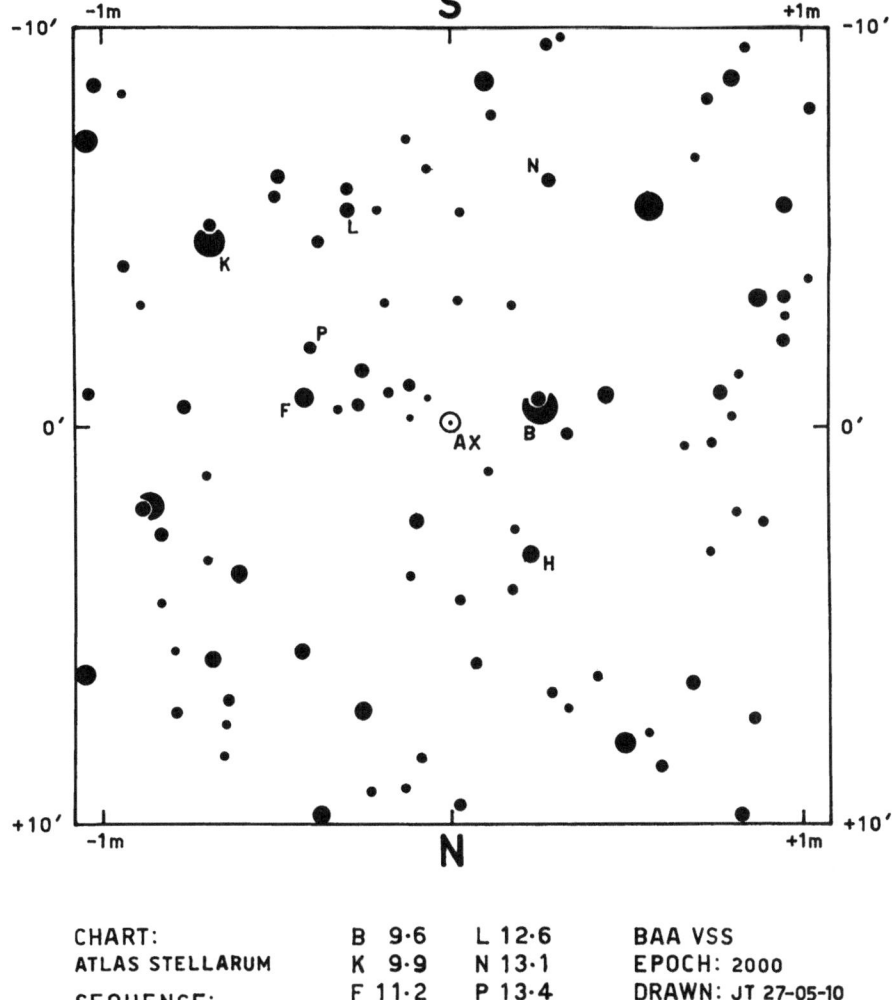

CHART: B 9·6   L 12·6   BAA VSS
ATLAS STELLARUM   K 9·9   N 13·1   EPOCH: 2000
                          F 11·2   P 13·4   DRAWN: JT 27-05-10
SEQUENCE: H 12·2   APPROVED: RDP
B & K TYCHO 2 VJ
OTHERS USNO

Fig. 14.25 AX Persei

056·02   1° FIELD INVERTED

RV TAURI   04h 47m 06·7s   +26° 10′ 46″   (2000)

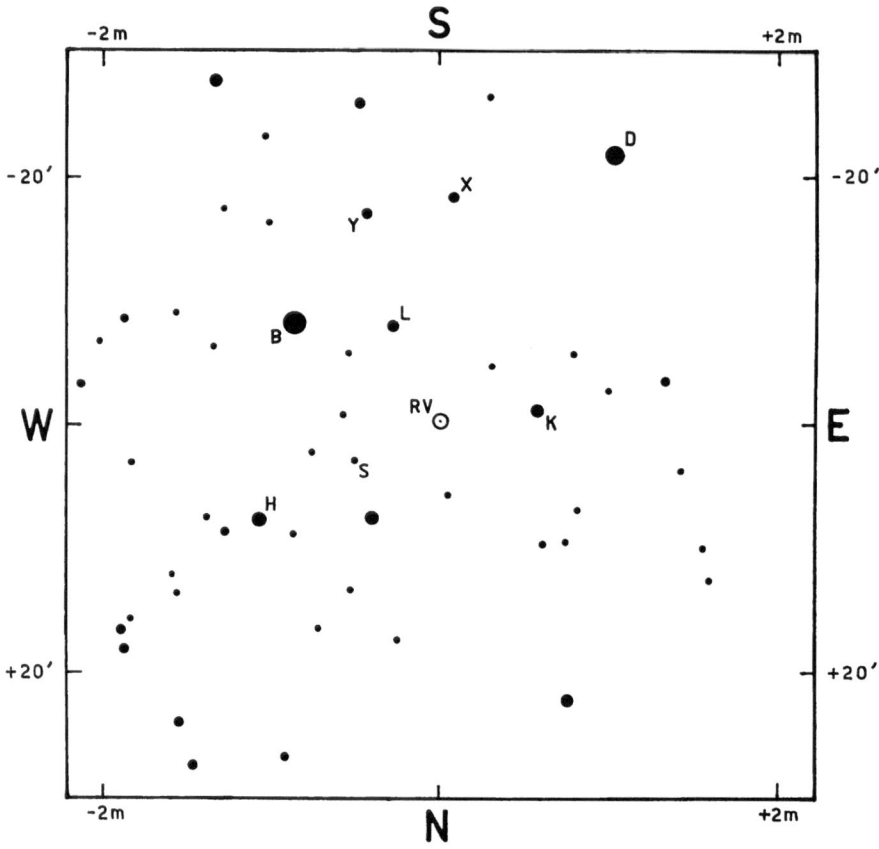

CHART:	B 7·7	L 11·1	BAA VSS
ATLAS STELLARUM	D 8·9	X 11·7	EPOCH: 2000
SEQUENCE:	H 10·0	Y 12·1	DRAWN: JT 09-08-12
B,D,H TYCHO 2 VJ	K 10·6	S 12·6	APPROVED: RDP
K HIPPARCOS VJ			
L,S,Y TASS			
X ASAS 3			

Fig. 14.26 RV Tauri

294·01   9° FIELD DIRECT

TU GEMINORUM   06h 10m 53·1s +26° 00' 53" (2000)
WY GEMINORUM   06h 11m 56·3s +23° 12' 25" (2000)
BU GEMINORUM   06h 12m 19·1s +22° 54' 31" (2000)
TV GEMINORUM   06h 11m 51·4s +21° 52' 06" (2000)

CHART:	A 5·8	Y 6·9	BAA VSS
ATLAS ECLIPTICALIS	D 6·3	Q 7·5	EPOCH: 2000
SEQUENCE:	F 6·9	S 7·7	DRAWN: JT 11-03-06
TYCHO 2 VJ	H 7·4	X 8·1	APPROVED: RDP
	K 7·8	Z 8·8	
	T 8·4		

Fig. 14.27 BU Geminorum

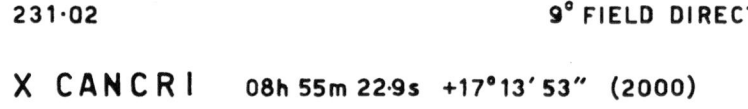

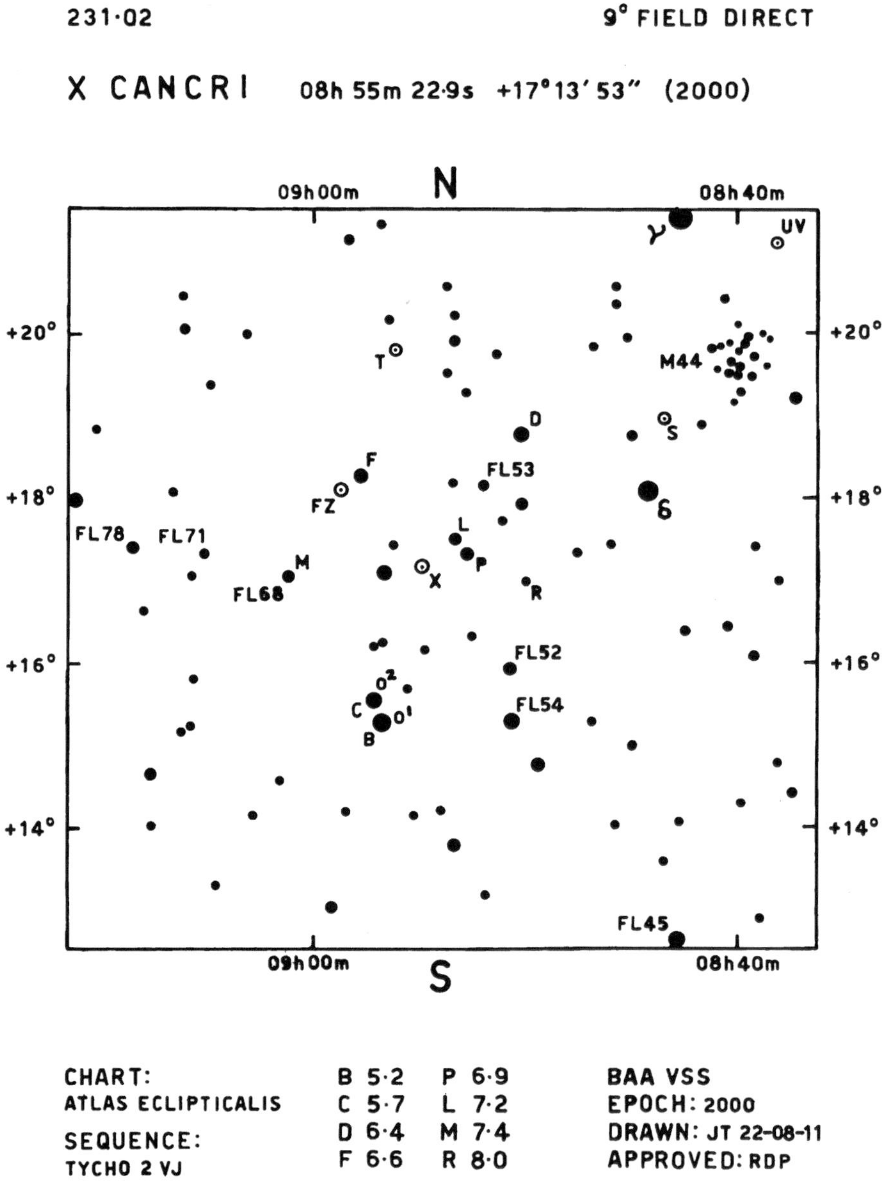
231·02
9° FIELD DIRECT
X CANCRI 08h 55m 22·9s +17°13′53″ (2000)
N
S
09h00m
08h40m
+20°
+18°
+16°
+14°
UV
M44
T
D
S
F
FZ
FL53
δ
FL78
FL71
L
M
P
X
R
FL68
FL52
O2
C
O1
B
FL54
FL45
CHART:
ATLAS ECLIPTICALIS
SEQUENCE:
TYCHO 2 VJ
B 5·2
C 5·7
D 6·4
F 6·6
P 6·9
L 7·2
M 7·4
R 8·0
BAA VSS
EPOCH: 2000
DRAWN: JT 22-08-11
APPROVED: RDP

Fig. 14.28 X Cancri

(Finder charts courtesy of the British Astronomical Association Variable Star Section)

R CORONAE BOR
Type: RCB
Magnitude range: 5.7 to 14.8
Period: –
RA: 15.48.34   Dec: 28.09.24

One of the most enigmatic variables in the sky, R Coronae Borealis was discovered by the great observer Edward Piggot in 1795. It later became, in 1935, relatively famous as the first star shown to have a different chemical composition than the Sun.

R Coronae Borealis is the prototype of the R Coronae Borealis class of variable stars. It is one of only two R Coronae Borealis variables bright enough to be seen with the naked eye, along with RY Sagittarii. For long periods of time, the star shows small variations of around a tenth of a magnitude with poorly defined periods, though these have been reported to be as little as 40 and 51 days. The star itself is thought to be a helium burning star much like the Sun will become, as the estimated mass of R Cor Bor is similar.

At irregular intervals a few years or decades apart R Coronae Borealis fades from its normal brightness near 6th magnitude for a period of months or sometimes years. There is no fixed minimum, but the star can become fainter than 15th magnitude in the visual range before it starts to return to maximum brightness almost immediately from its minimum, although occasionally this is interrupted by another fade. The cause of this behavior is believed to be a result of the convective processes of such stars, dredging up carbon from the interior before the carbon undergoes condensation in the upper atmosphere of the star, resulting in a rapid drop in luminosity and the light being blocked. As the material is dispersed by radiation pressure, the light slowly returns to normal brightness.

In August 2007, R Coronae Borealis began a fade below its usual minimum, falling to 14th magnitude in 33 days before fading again slowly to below 15th magnitude in June 2009. It then began a very slow rise, reaching 12th magnitude in late 2011. R Coronae Borealis then faded again to near 15th magnitude, and by August 2014 it had been below 10th magnitude for 7 years.

In late 2014, R Coronae Borealis brightened quickly to 7th magnitude but then began to fade again, and the last few observations in late 2017 show that the star has now been below normal brightness for over a decade. It could rise back to normal at any time and is a wonderful star to follow.

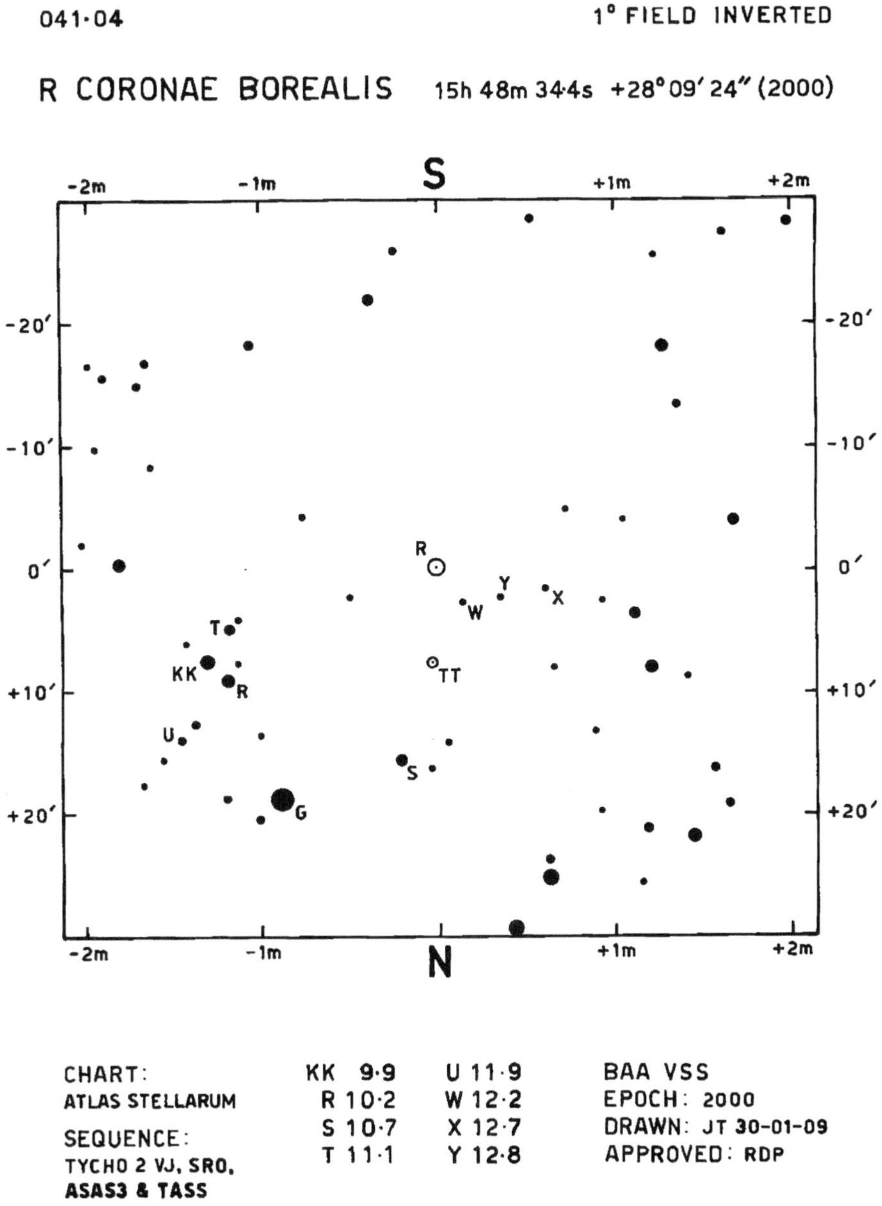
041·04
1° FIELD INVERTED
R CORONAE BOREALIS 15h 48m 34·4s +28°09′24″ (2000)
S
N
-2m
-1m
+1m
+2m
-20′
-10′
0′
+10′
+20′
R
TT
KK
T
U
G
W
X
Y
CHART:
ATLAS STELLARUM
SEQUENCE:
TYCHO 2 VJ, SRO,
ASAS3 & TASS
KK 9·9
R 10·2
S 10·7
T 11·1
U 11·9
W 12·2
X 12·7
Y 12·8
BAA VSS
EPOCH: 2000
DRAWN: JT 30-01-09
APPROVED: RDP

Fig. 14.29 R Coronae Borealis

(Finder chart courtesy of the British Astronomical Association Variable Star Section)

R HYDRAE
Type: M
Magnitude range: 3.5 to 10.9
Period: 389d
RA: 13.29.42   Dec: −23.16.53

R Hydra is a Mira-type long-period variable in the southern constellation of the water snake and is a lovely object for a beginner to follow, as large aperture telescopes are not needed, and a good light curve can be obtained due to its broad magnitude range. R Hydrae appears to be about 400 light years from Earth, and the Spitzer space telescope reveals a bow shock ahead of the star's movement as its magnetic field interacts with the interstellar medium. At maximum luminosity it is an easy naked-eye object and has a slowly changing period that is very slowly decreasing.

049·03   9° FIELD DIRECT

R HYDRAE   13h 29m 42·8s   −23° 16′ 53″ (2000)

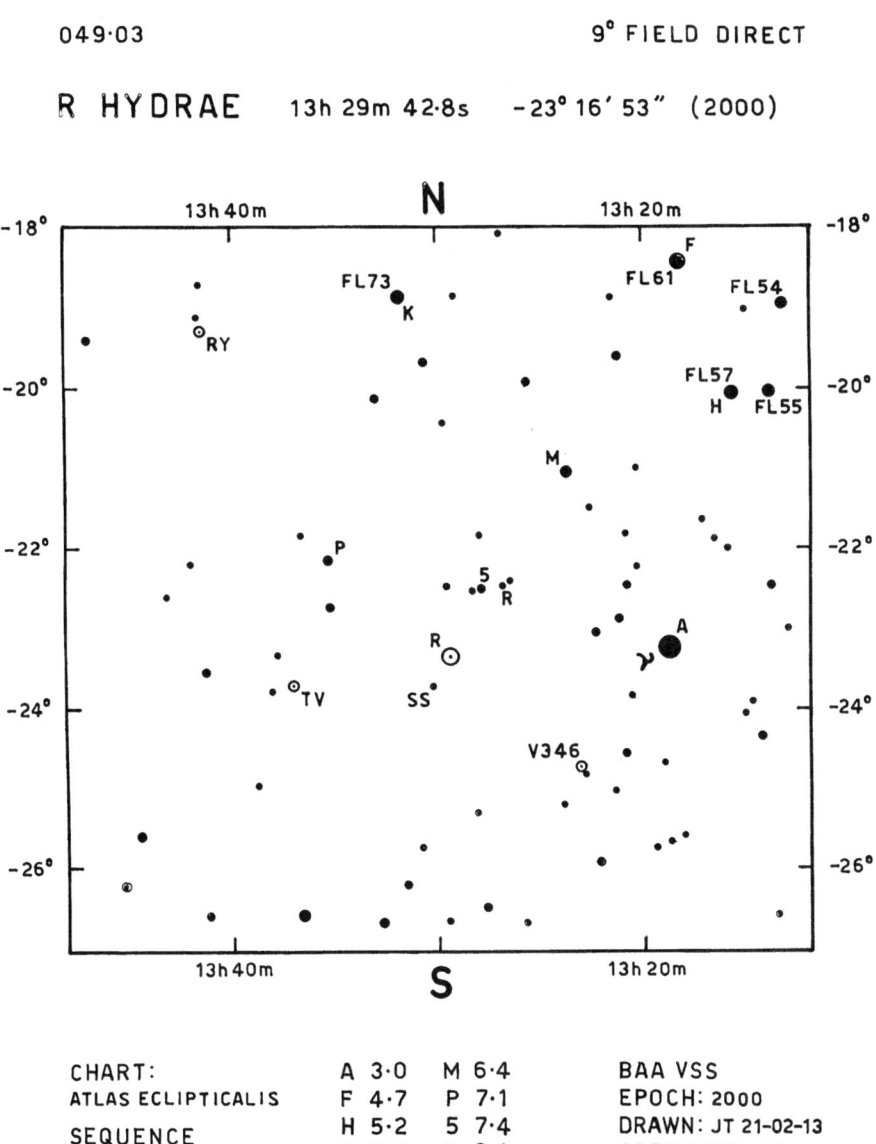

CHART:			BAA VSS
ATLAS ECLIPTICALIS	A 3·0	M 6·4	EPOCH: 2000
	F 4·7	P 7·1	DRAWN: JT 21-02-13
SEQUENCE	H 5·2	5 7·4	APPROVED: RDP
HIPPARCOS VJ	K 6·0	R 8·1	

Fig. 14.30 R Hydrae

(Finder chart courtesy of the British Astronomical Association Variable Star Section)

RX BOOTES
Type: SRB
Magnitude range: 8.6 to 11.3
Period: 160d
RA: 14.21.57   Dec: 25.55.8

This semi-regular red giant star is a good variable star for observers, as its magnitude range lies well within the field of even smaller telescopes. Although catalogued as a SRB, the star actually has a fairly regular underlying period of 160 days and a secondary period of 302 days. This Mira-type star is surrounded by a cloud of gas and dust in which various oxides have been analyzed spectroscopically, and from radio astronomy the cloud is a source of water MASER emission, too.

219·02   9° FIELD DIRECT

RX BOOTIS   14h 24m 11·6s   +25° 42' 14"   (2000)

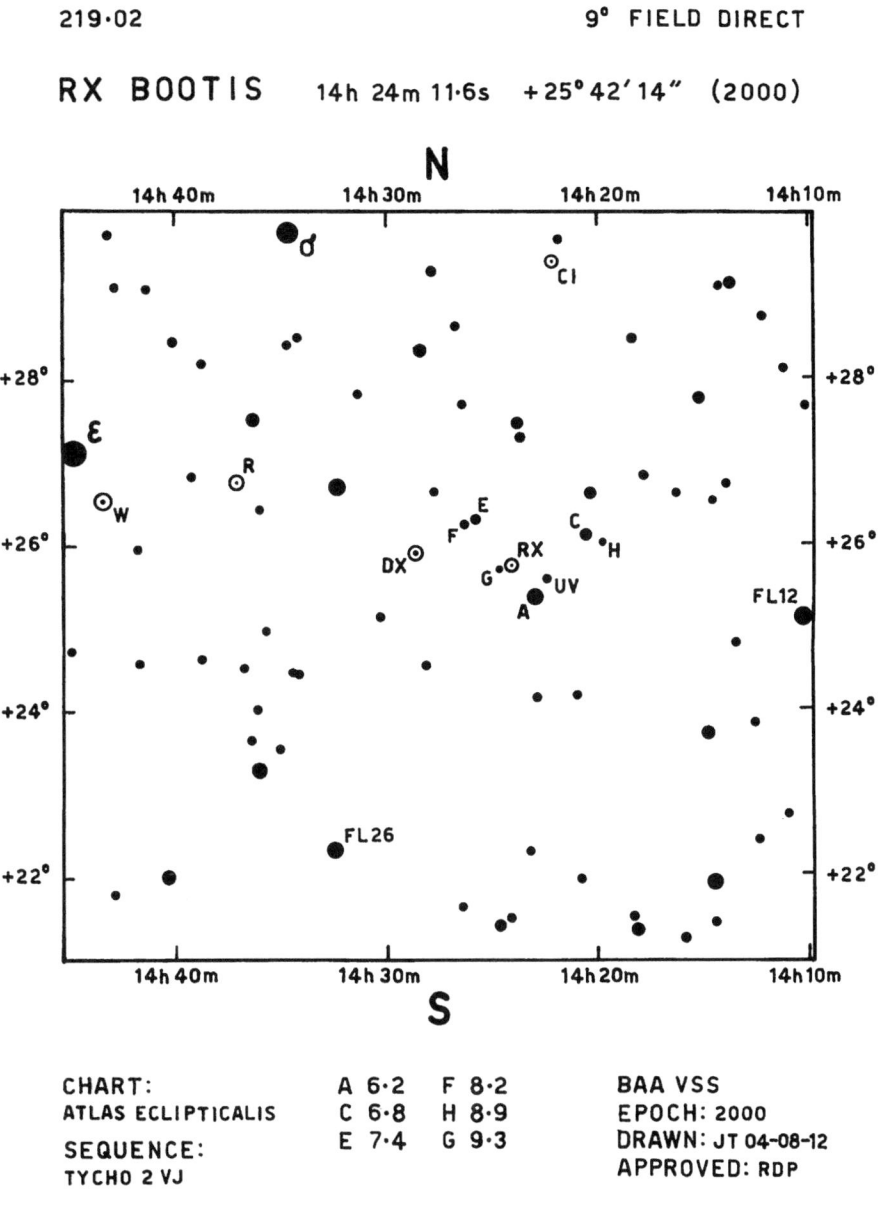

CHART:	A 6·2	F 8·2	BAA VSS
ATLAS ECLIPTICALIS	C 6·8	H 8·9	EPOCH: 2000
SEQUENCE:	E 7·4	G 9·3	DRAWN: JT 04-08-12
TYCHO 2 VJ			APPROVED: RDP

Fig. 14.31 RX Bootes

(Finder chart courtesy of the British Astronomical Association Variable Star Section)

T CORONAE BOR
Type: NR
Magnitude range: 10.8 to 2
Period: 29,000d
RA: 15.59.30   Dec: 25.55.12

The prototype of a class of recurrent novae, T Coronae Borealis has a typical period of about 80 years or so and has been seen several times in the last two centuries. Originally thought to be a 10th magnitude star, it was caught in outburst by the Irish astronomer John Birmingham in 1866, where it attained magnitude 2 before fading away. In 1946 the star again was visible to the naked eye at 3rd magnitude. It has been reported that the star has been brightening since 2016, and such an increase in brightness was reported before its outburst in 1946, so who knows? It may well go up to naked-eye magnitude soon, so it's worth scanning the constellation for!

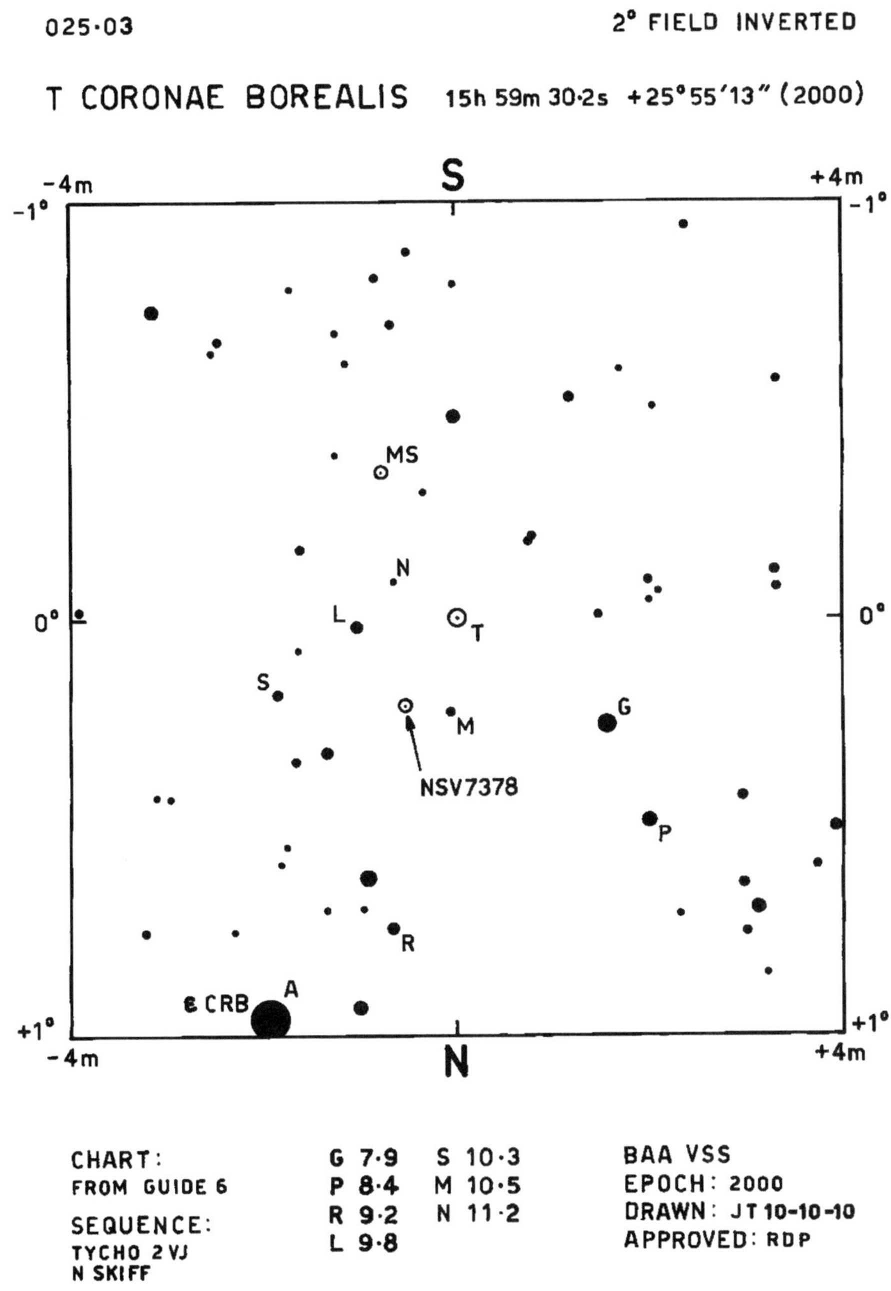
025·03
2° FIELD INVERTED
T CORONAE BOREALIS 15h 59m 30·2s +25°55′13″ (2000)
S
N
-4m
+4m
-1°
0°
+1°
MS
N
L
T
S
M
G
NSV7378
P
R
ε CRB
A
CHART:
FROM GUIDE 6
SEQUENCE:
TYCHO 2 VJ
N SKIFF
G 7·9
P 8·4
R 9·2
L 9·8
S 10·3
M 10·5
N 11·2
BAA VSS
EPOCH: 2000
DRAWN: JT 10-10-10
APPROVED: RDP

Fig. 14.32 T Coronae Borealis

(Finder chart courtesy of the British Astronomical Association Variable Star Section)

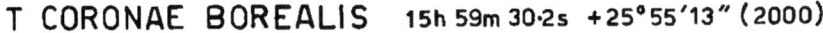

RS CANES VENATICORUM
Type: EA
Magnitude range: 7.9 to 9.1
Period: 4.79d
RA: 13.10.36   Dec: 35.56.05

RS Canes Venaticorum is an eclipsing or rotating Algol-type star of the AR Lacertae subtype, which means that its physical size changes over time due to the close proximity of the two stars in the binary system. The stars regularly undergo eclipses, and the dynamics of the interchange between the close orbiting pair can be gauged by the fact that this star is also an X-ray variable. RS Canes Venaticorum has a good range of magnitudes for easy amateur study and a very regular period. Timings of its eclipses can be obtained from the BAA or the AAVSO

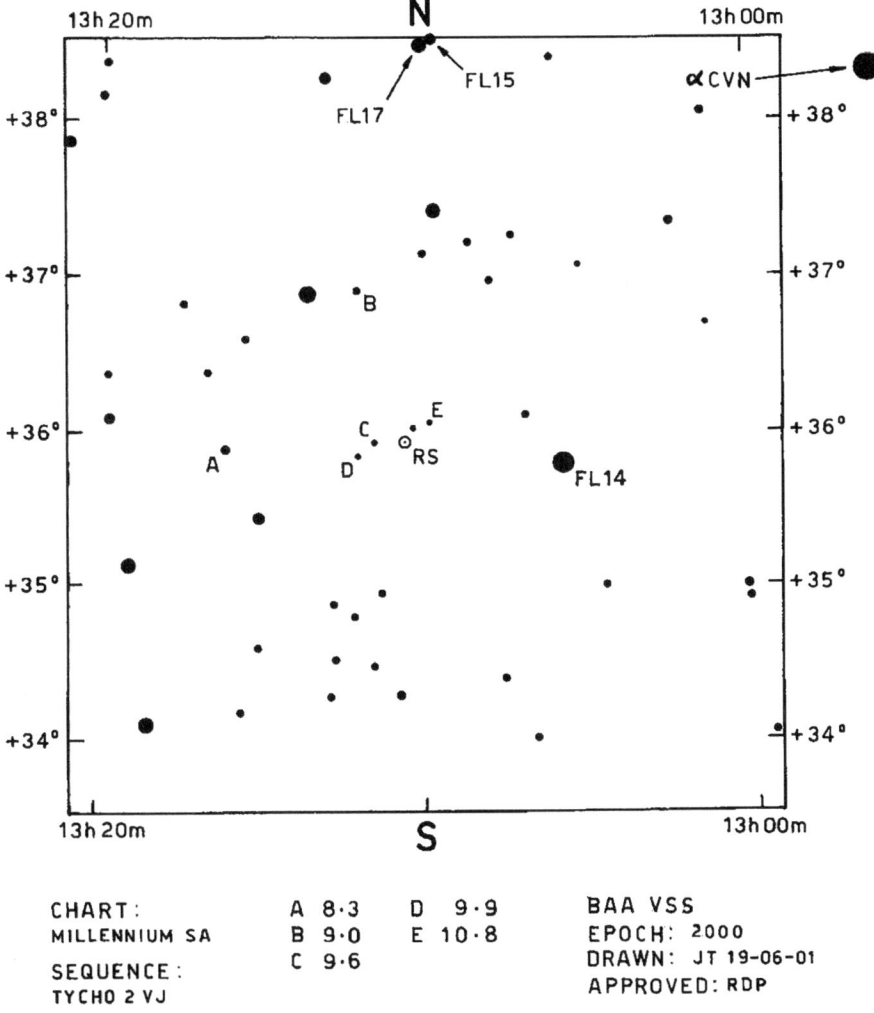

253·01   5° FIELD DIRECT

RS CANUM VENATICORUM   13h 10m 36·9s +35° 56′ 05″ (2000)

CHART: A 8·3   D 9·9   BAA VSS
MILLENNIUM SA   B 9·0   E 10·8   EPOCH: 2000
                C 9·6   DRAWN: JT 19-06-01
SEQUENCE: APPROVED: RDP
TYCHO 2 VJ

Fig. 14.33 RS Canes Venaticorum

(Finder chart courtesy of the British Astronomical Association Variable Star Section)

AC HERCULIS
Type: RVA
Magnitude range: 6.9 to 9
Period: 75d
RA: 18.13.16   Dec: 21.52.00

AC Her is a very clear example of a common type of RV Tauri light curve where the maximum following a deep minimum is brighter than the maximum following a shallow minimum. In each period of 75 days it has two maxima and two minima. AC Herculis is also a binary star, although the secondary can only be detected by its effect on the radial velocity of the primary. The invisible secondary is more massive than the supergiant primary, so the primary moves at relatively high velocity in its 3.3-year orbit. The two stars are also surrounded by a dusty disc filling a large region between 34 and 200 AU across.

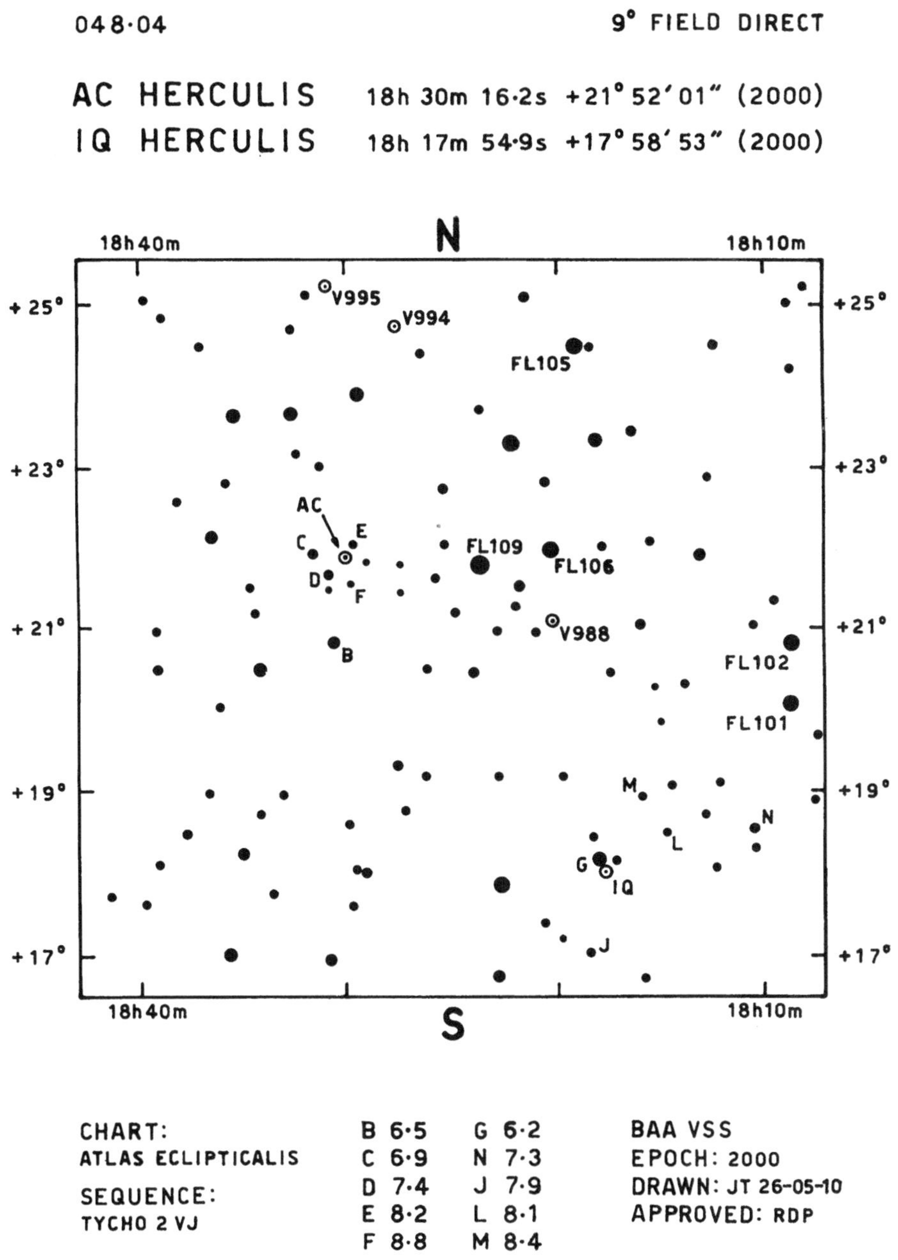

AM HERCULIS
Type: AM
Magnitude range: 12.3 to 15.7
Period: 0.12d
RA: 18.16.13   Dec: 49.52.04

This little red dwarf is a faint system that lies on the boundary of visibility in a 300-mm telescope, but its rise to brightness can be rapid before it fades away equally rapidly. It is a typical cataclysmic variable star of the type known as polars, due to their intense magnetic fields that may alter the way gas is accreted from a companion. Usually, such gas falls through the inner Lagrange point onto the equator of the companion, but the magnetic fields of such polars, up to several hundred million Gauss on occasion, channel the gas around the field lines, onto the magnetic pole of the white dwarf. The star was first catalogued in 1923 by the astronomer Max Wolf. By the end of the 20th century it had been revealed that AM Herculis is highly magnetized, and its light output is polarized as a result.

This is a faint object, usually below the observable threshold in most amateur telescopes, but the field is well worth scanning just in case the star has brightened and can be recognized.

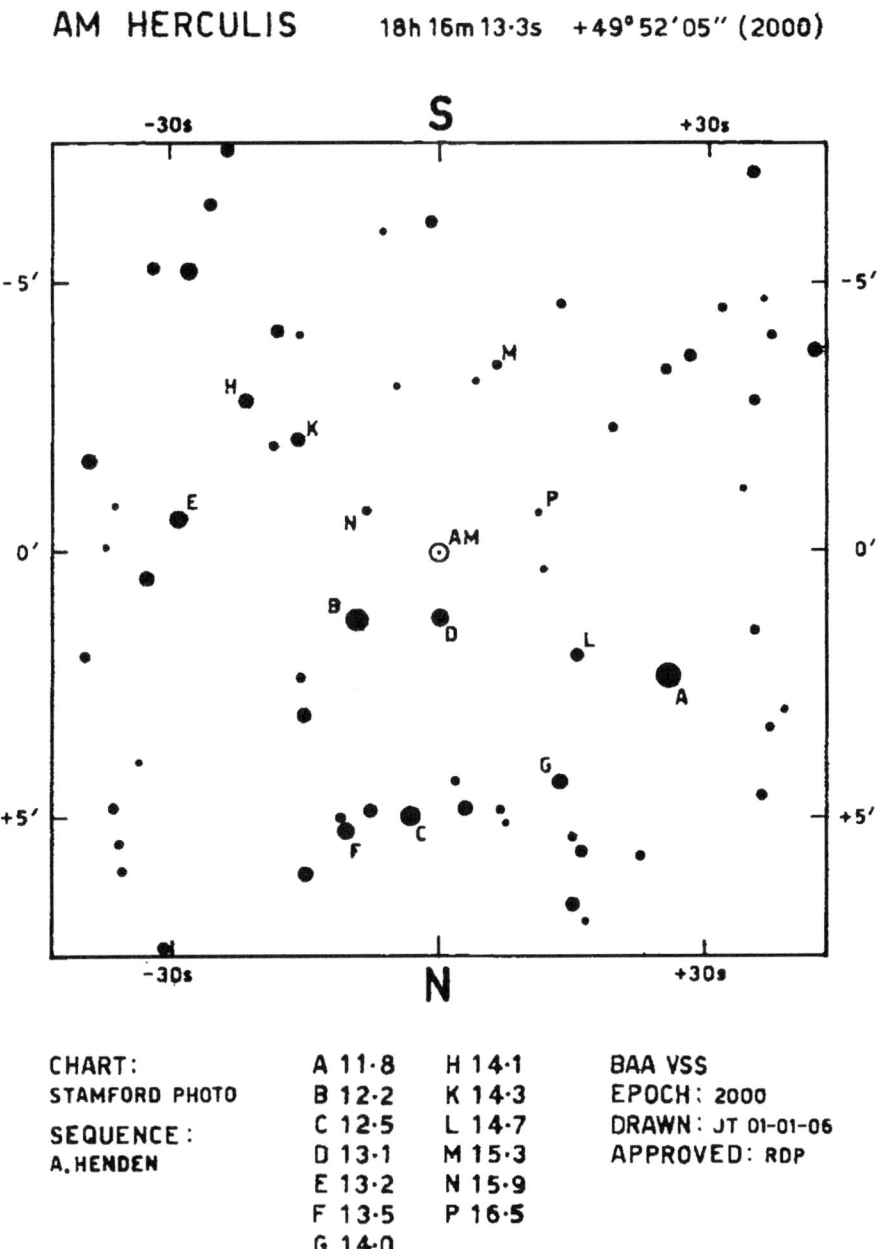

Fig. 14.35 AM Herculis

(Finder chart courtesy of the British Astronomical Association Variable Star Section)

For observers wishing to add to their observing list, a few suggestions from the BAA Variable Star Section are listed here along with their finder charts included.

Star	Type	Range	Period	Frequency
V CVn	SRa	6.5–8.6	192d	5d-7d
U Boo	SRb	9.8–13.0	201d	5d-7d
V Boo	SRa	7.0–12.0	258d	5d-7d
X Leo	UGSS	11.1–16.5	17d	Nightly

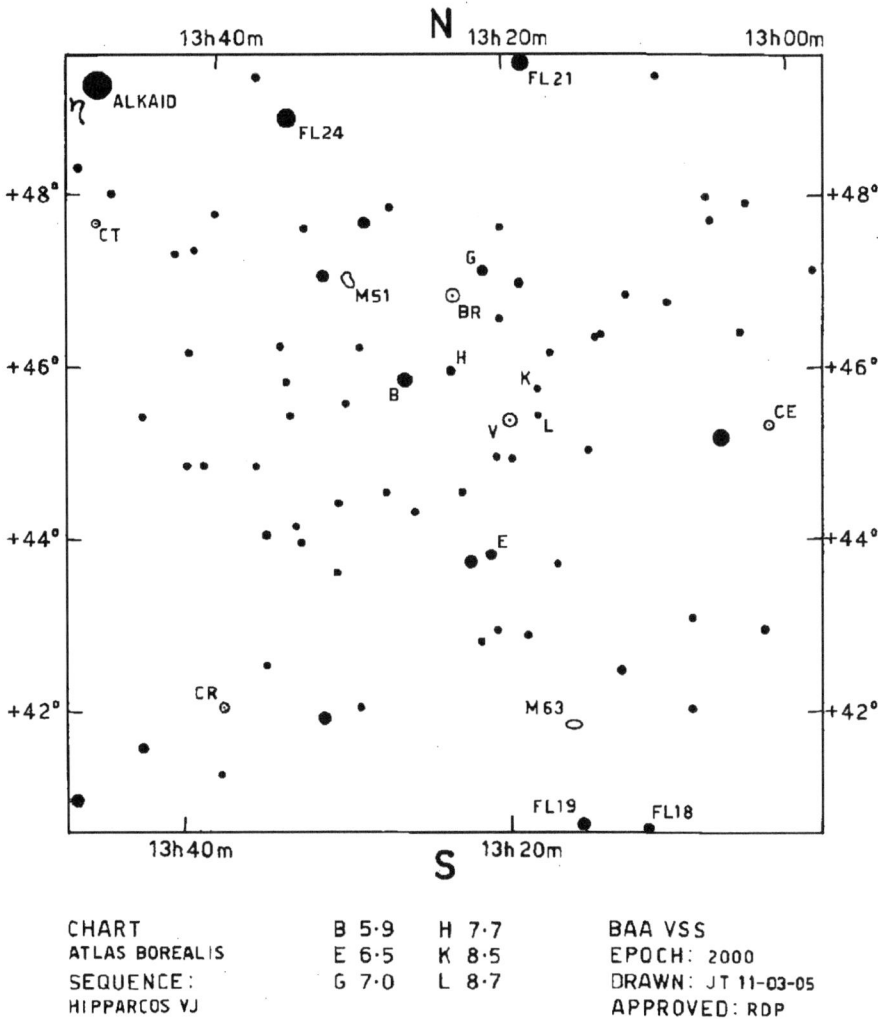

214·02   9° FIELD DIRECT

V CANUM VENATICORUM   13h 19m 27·8s +45° 31′ 38″ (2000)

CHART   B 5·9   H 7·7   BAA VSS
ATLAS BOREALIS   E 6·5   K 8·5   EPOCH: 2000
SEQUENCE: G 7·0   L 8·7   DRAWN: JT 11-03-05
HIPPARCOS VJ   APPROVED: RDP

Fig. 14.36 V Can Ven

036·02   1° FIELD INVERTED

U BOOTIS   14h 54m 20·0s +17°41′44″ (2000)

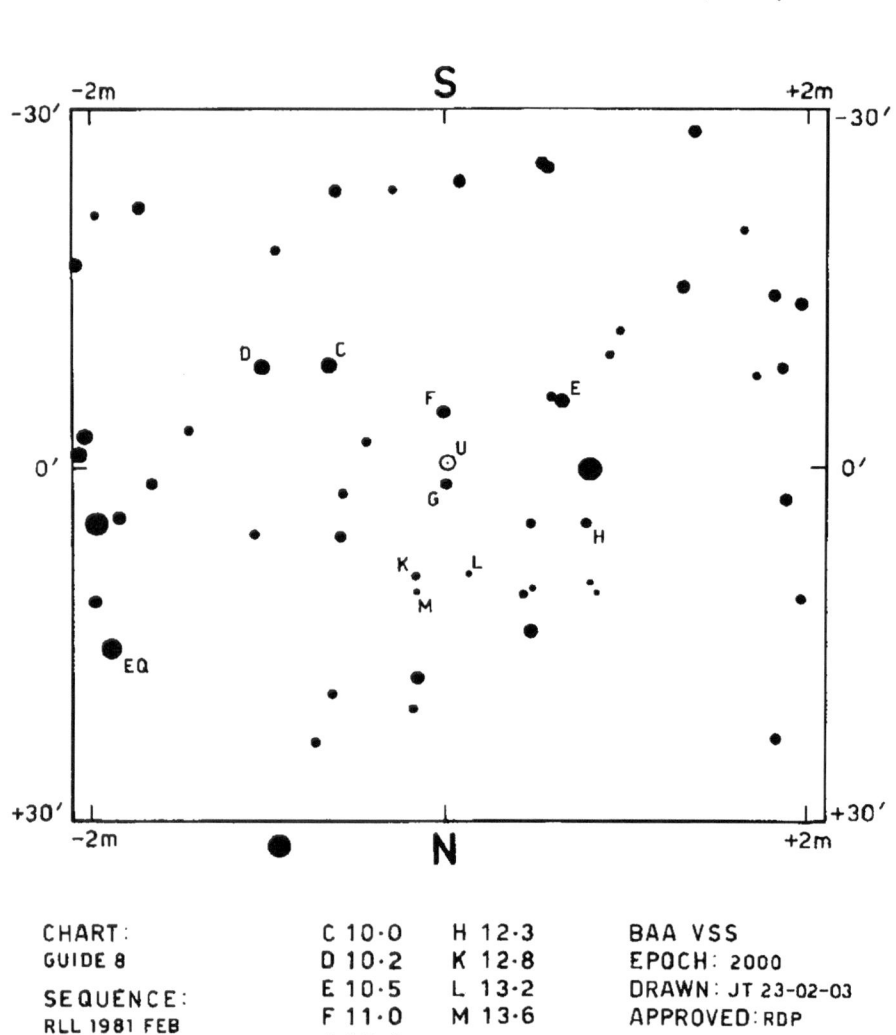

CHART:	C 10·0	H 12·3	BAA VSS
GUIDE 8	D 10·2	K 12·8	EPOCH: 2000
SEQUENCE:	E 10·5	L 13·2	DRAWN: JT 23-02-03
RLL 1981 FEB	F 11·0	M 13·6	APPROVED: RDP
	G 12·0		

Fig. 14.37 U Bootis

037·02 3° FIELD INVERTED

V BOOTIS 14h 29m 45·3s +38° 51′ 41″ (2000)

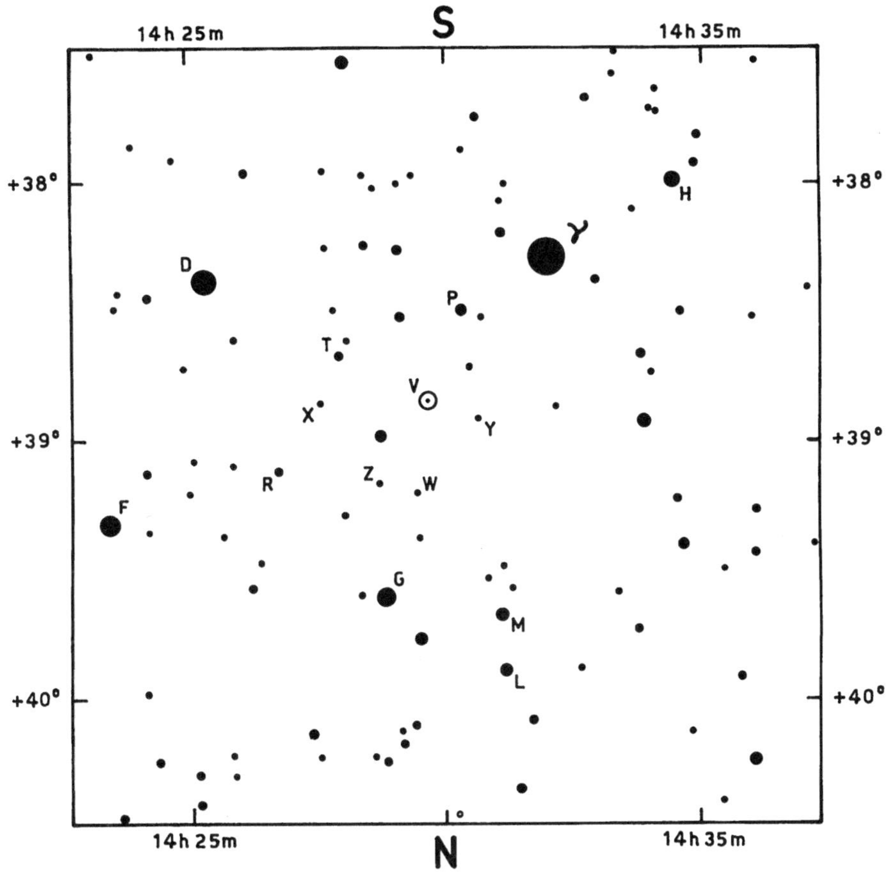

CHART:
MILLENNIUM SA

SEQUENCE:
D-T TYCHO 2 VJ
W-Z RDP

D	6·3	R	9·6
F	7·2	T	9·9
G	7·5	W	10·6
H	8·1	X	11·0
L	8·7	Y	11·4
M	8·8	Z	11·7
P	9·2		

BAA VSS
EPOCH: 2000
DRAWN: JT 03-08-12
APPROVED: RDP

Fig. 14.38 V Bootis

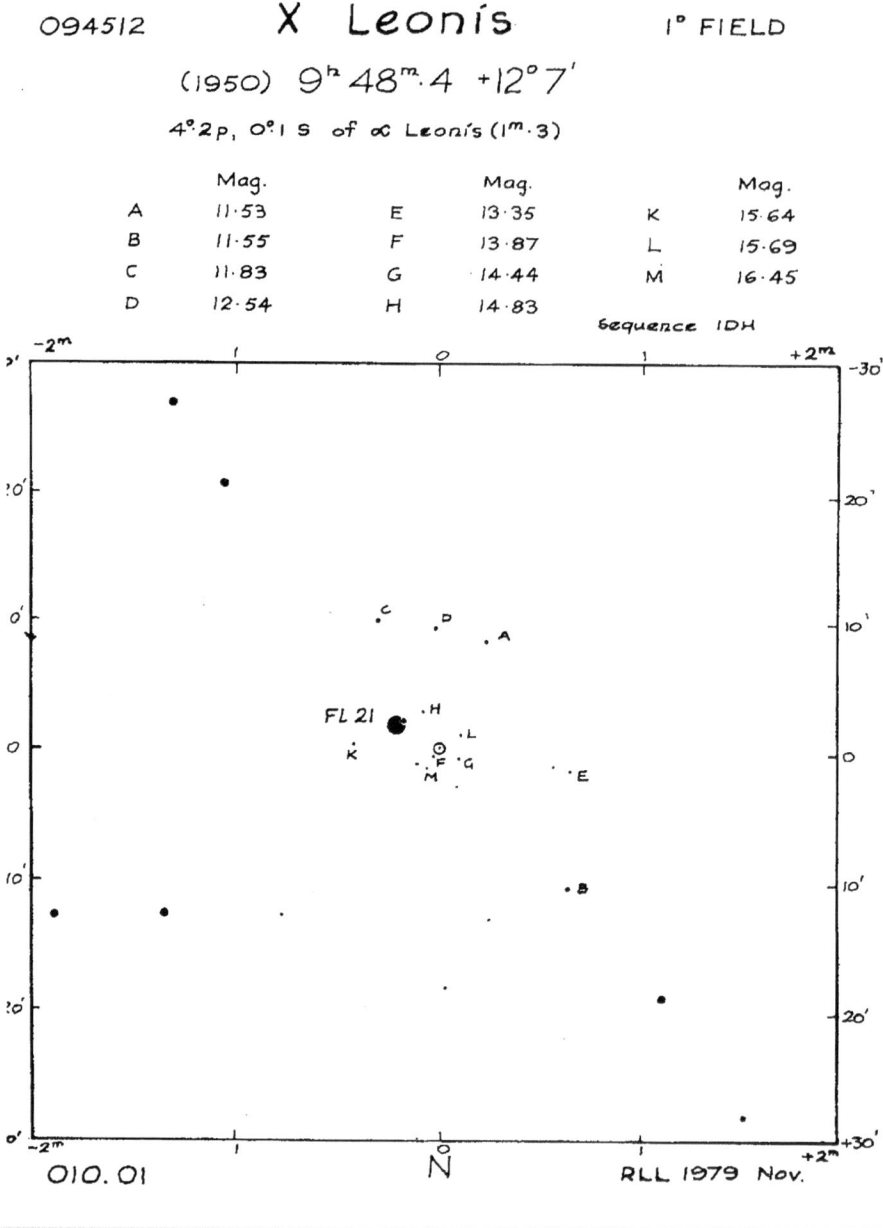

Fig. 14.39 X Leonis

(Finder chart courtesy of the British Astronomical Association Variable Star Section)

FG SAGITTAE
Type: Unique
Magnitude range: 9.5 to 13.6
Period: –
RA: 20.11.56   Dec: 20.20.04

One of the strangest variables in the sky, the faint star FG Sagittae was first noticed to be variable during WW2 in 1943, and its spectra changed from a blue star, first thought to be a supergiant, through to G-type and finally an orange K-type before settling to become the A-type star we see today. This variability is due to the star being the central one of the planetary nebulae Henize 1–5 and as such is becoming a white dwarf.

The nebula itself is a very difficult object to see visually, as the gas still surrounds the star quite closely, given the distance to the object is in excess of 8,000 light years. The star also shows fading behavior, like an RCB type star, probably due to the swirling effects of ejected gas from the thermal pulsations causing the planetary nebula. It is possible that FG Sagittae was caught during its last thermal pulsation to become the white dwarf we see today. Its fluctuations in light output are still present, so this rather interesting object has been caught in the act of dying.

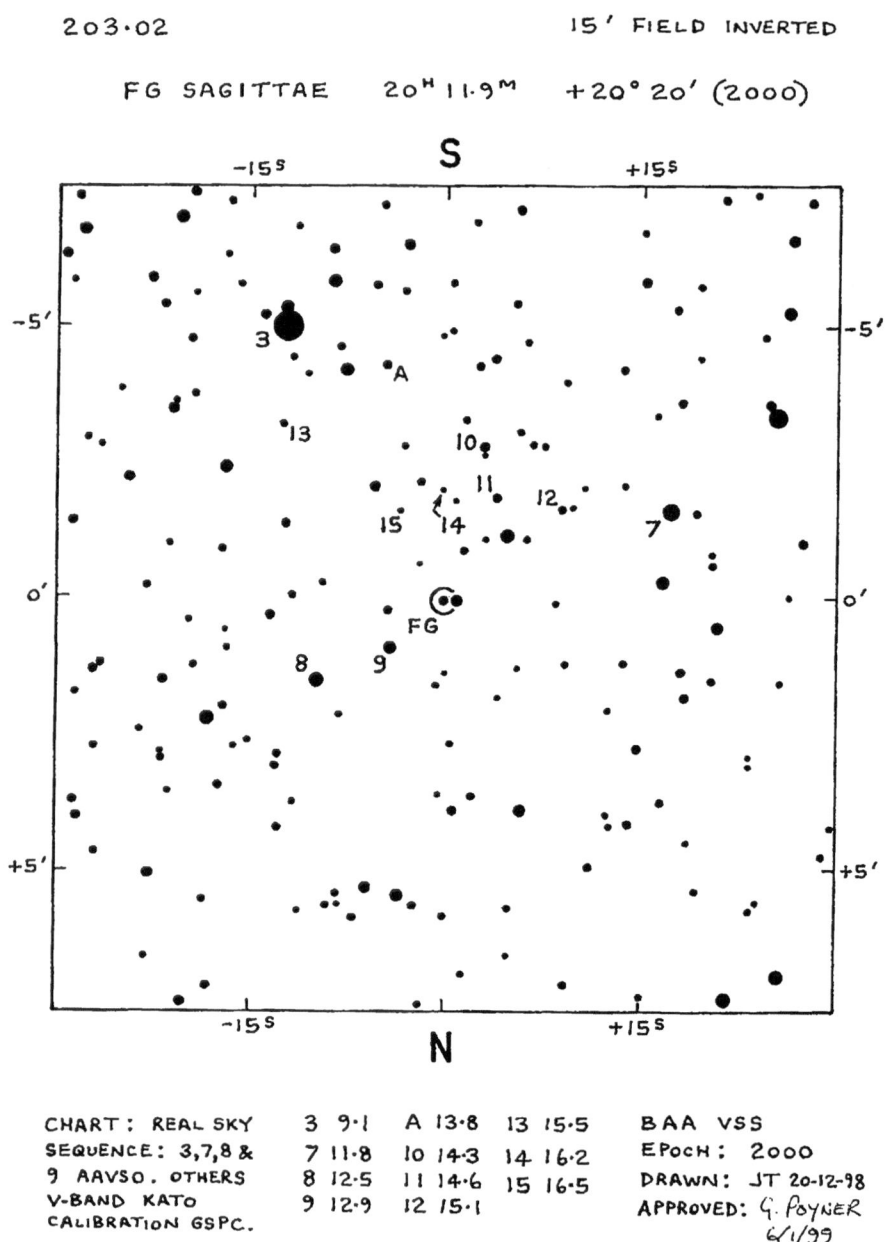

Fig. 14.40 FG Sagittae

(Finder chart courtesy of the British Astronomical Association Variable Star Section)

R AQUILAE
Type: M
Magnitude range: 5.3 to 12
Period: 277d
RA: 19.06/22   Dec: 08.13.48

R Aquilae is a typical long-period variable of the Mira-type with a large magnitude range and is thus a great object for amateurs to concentrate on. The period of variability has been declining from the 300+ first noticed in the 19th century, and now, the period as indicated above, is 277 days. Astronomers are a little perplexed by this variability, but it may be due to mass outflow or ejection ridding the star of matter and resulting in a shortened pulsation period.

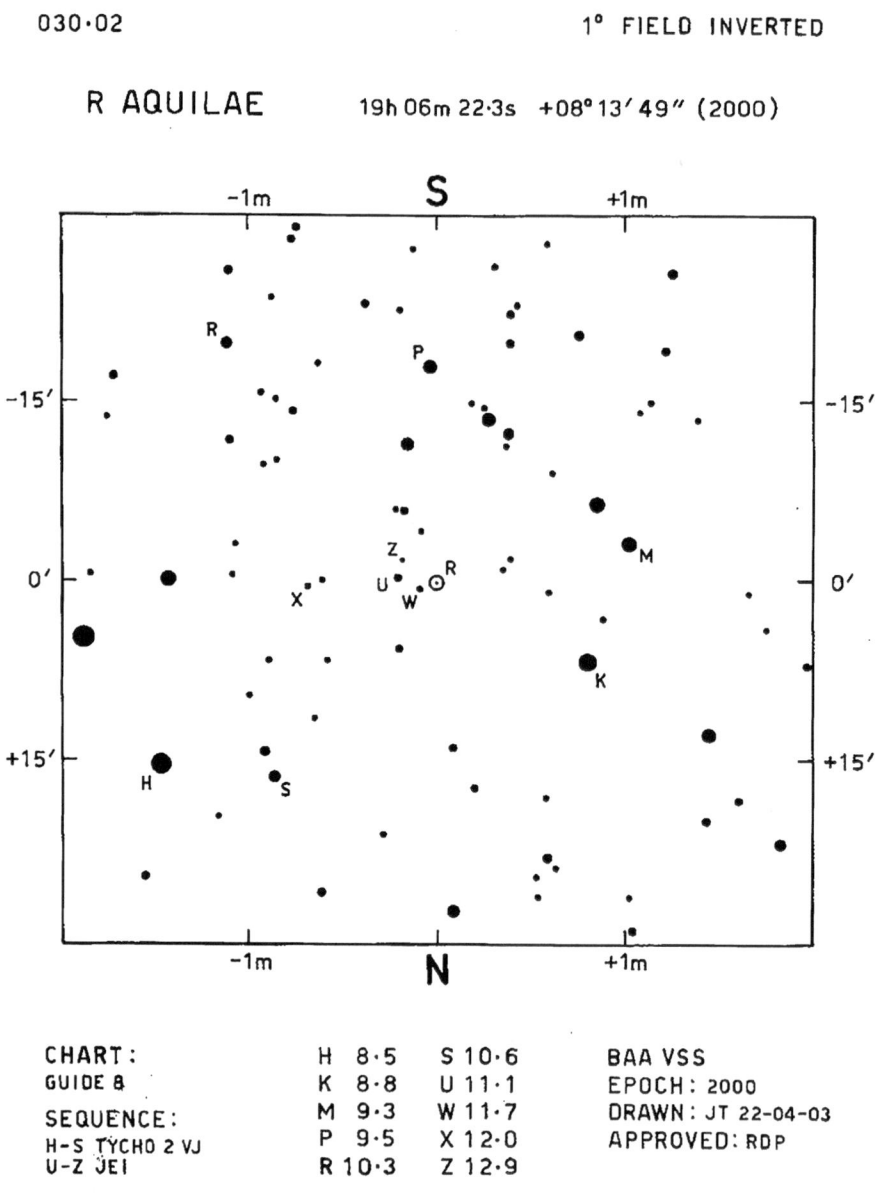

030·02   1° FIELD INVERTED

R AQUILAE   19h 06m 22·3s   +08° 13′ 49″ (2000)

CHART:	H 8·5	S 10·6	BAA VSS
GUIDE 8	K 8·8	U 11·1	EPOCH: 2000
SEQUENCE:	M 9·3	W 11·7	DRAWN: JT 22-04-03
H-S TYCHO 2 VJ	P 9·5	X 12·0	APPROVED: RDP
U-Z JEI	R 10·3	Z 12·9	

Fig. 14.41 R Aquilae

(Finder chart courtesy of the British Astronomical Association Variable Star Section)

R SCUTI
Type: RVA
Magnitude range: 4.2 to 8.6
Period: 146.5d
RA: 18.47.28   Dec: −05.42.18

One of the variable stars discovered by Edward Piggot in 1795, R Scuti is located to the northwest of the beautiful star cluster Messier 11, "the wild duck." Given its longevity, it is hardly surprising that over 15,000 observations of this star exist going back almost two centuries, making R Scuti one of the most studied stars in history.

The star is a RV Tauri-type variable, a pulsating yellow supergiant. Such variables often have somewhat irregular light curves, both in amplitude and period, but R Scuti is extreme. It has one of the longest periods known for an RV Tau variable, and the light curve has a number of unusual features: occasional extreme minima, intermittent standstills with only small erratic variation that may last for years and periods of chaotic brightness changes. In deep minima, much of the spectrum corresponds to an early K supergiant, but the spectrum also reveals titanium oxide absorption bands more typical of an M-type star. If R Scuti is a post-AGB star, then it would show measurable changes in its temperature and period over the time that R Scuti has been closely observed. Instead, astronomers discovered that the star has a relatively low mass loss rate with an extended cool atmosphere. R Scuti also has a fairly constant temperature and period of variation. It is possible that R Scuti is still a thermally pulsing AGB star, which appears consistent with calculated levels of mass loss that are being observed from this star.

026·04   8° FIELD DIRECT

V AQUILAE   19h 04m 24·2s   −05° 41′ 05″   (2000)

R SCUTI   18h 47m 29·0s   −05° 42′ 18″   (2000)

S SCUTI   18h 50m 20·0s   −07° 54′ 27″   (2000)

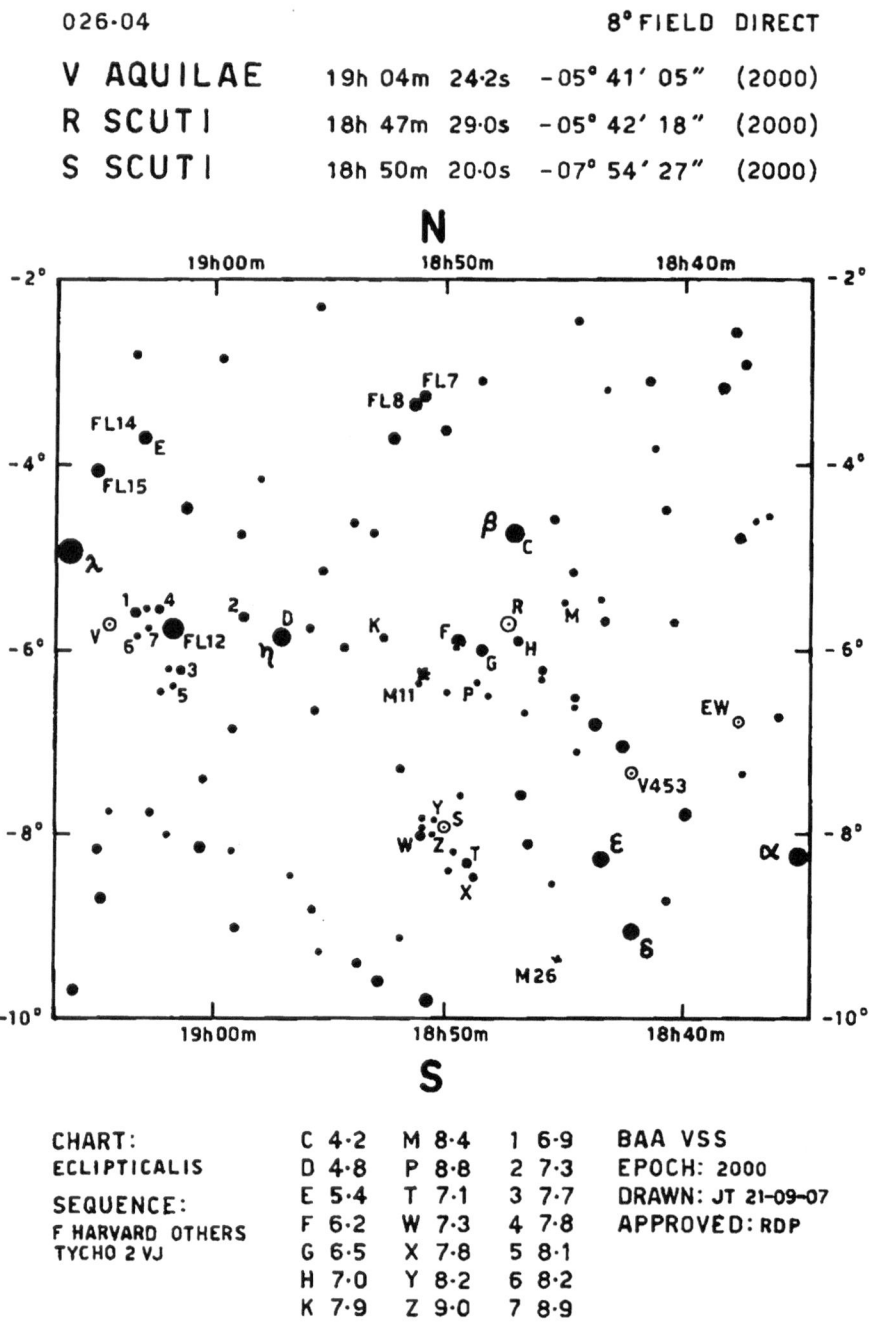

CHART:	C 4·2	M 8·4	1 6·9	BAA VSS
ECLIPTICALIS	D 4·8	P 8·8	2 7·3	EPOCH: 2000
SEQUENCE:	E 5·4	T 7·1	3 7·7	DRAWN: JT 21–09–07
F HARVARD OTHERS	F 6·2	W 7·3	4 7·8	APPROVED: RDP
TYCHO 2 VJ	G 6·5	X 7·8	5 8·1	
	H 7·0	Y 8·2	6 8·2	
	K 7·9	Z 9·0	7 8·9	

Fig. 14.42 R Scuti

(Finder chart courtesy of the British Astronomical Association Variable Star Section)

RS OPHIUCHI
Type: NR
Magnitude range: 4.3 to 12.5
Period: –
RA: 17.50.13   Dec: –06.42.28

RS Ophiuchi is a recurrent novae system consisting of a red giant primary star and an orbiting white dwarf. Usually the star lies quiescent at a faint magnitude 12.5 but erupts irregularly and suddenly. Astronomers have observed this star at maximum eruption in 1898, 1933, 1958, 1967, 1985 and again in 2006, where the object reaches about magnitude 5 on average. These recurrent novae are produced by a white dwarf star and a red giant in a binary system. About every 20 years, enough material from the red giant builds up on the surface of the white dwarf to produce a huge thermonuclear explosion. In the meantime, the white dwarf orbits close to the red giant, pulling material off into an accretion disc, which then settles onto the white dwarf surface until pressure and heat buildup to catastrophic proportions, causing the material to fuse and explode.

Although the star is quite faint at its usual magnitude, it can be seen in apertures of 20 cm and above. Given its semi-regular nature, it is always worth checking the field for this wonderful yet rare little star.

024·02   1° FIELD INVERTED

RS OPHIUCHI   17h 50m 13·2s   −06°42′28″ (2000)

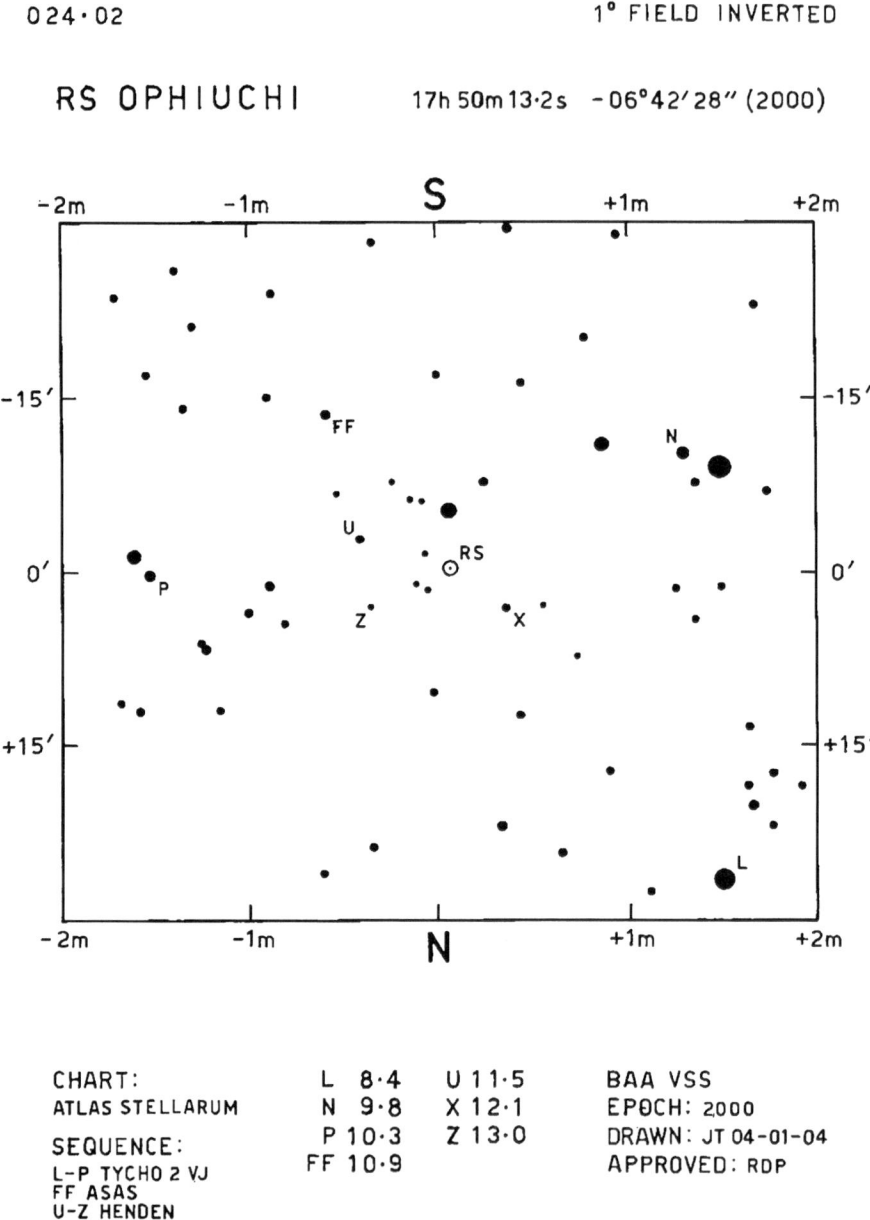

CHART:	L 8·4	U 11·5	BAA VSS
ATLAS STELLARUM	N 9·8	X 12·1	EPOCH: 2000
SEQUENCE:	P 10·3	Z 13·0	DRAWN: JT 04-01-04
L-P TYCHO 2 VJ	FF 10·9		APPROVED: RDP
FF ASAS			
U-Z HENDEN			

Fig. 14.43 RS Ophiuchi

(Finder chart courtesy of the British Astronomical Association Variable Star Section)

SS CYGNI
Type: UGSS
Magnitude range: 7.7 to 12.4
Period: 49.5d
RA: 21.42.42   Dec: 43.35.09

SS Cygni has been called the prototype dwarf nova and is a U Gemiorum-type cataclysmic variable undergoing frequent and regular brightness outbursts every 7 or 8 weeks or so. However, the maxima of SS Cygni stars are typically brighter than those of U Gem, rising from 12th magnitude to 8th magnitude for, typically, 1 to 2 days.

Like all other cataclysmic variables, SS Cygni consists of a close binary system with one of the components being a red dwarf type star, cooler than the Sun, while the other is a white dwarf. Studies suggest that the stars in the SS Cygni system are separated by only 100,000 miles or less; the stars are so close that they complete their orbital revolution in slightly over 6.5 hours. The magnitude range is typically large for cataclysmic variables, and the star can be seen in modest telescopes.

Astronomically speaking, SS Cygni is also fairly close by, at about 400 light years or so. Typical of red dwarfs, the stars are invisible to the naked eye unless the system goes into nova mode, and even then the maximum still requires a telescope to see it.

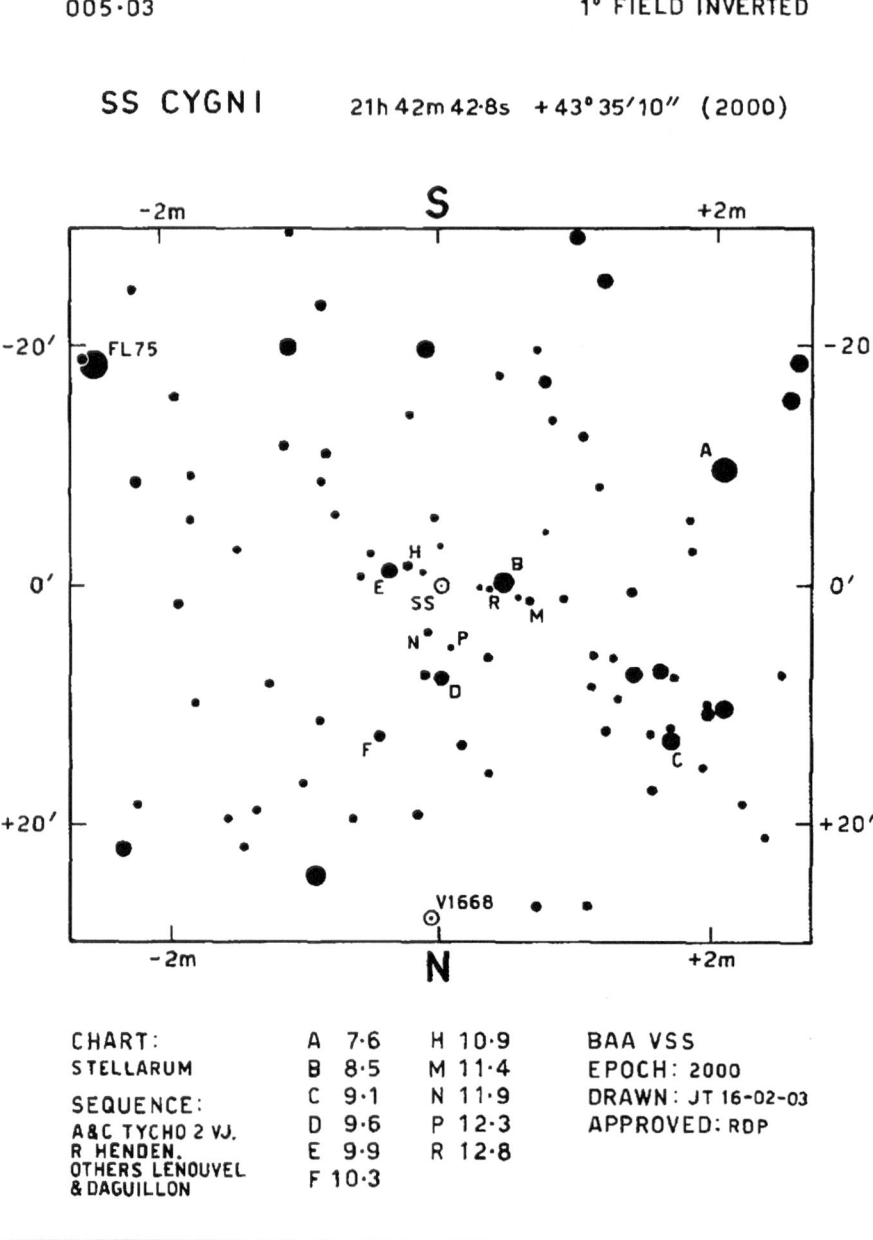

Fig. 14.44 SS Cygni

(Finder chart courtesy of the British Astronomical Association Variable Star Section)

WZ SAGITTAE
Type: UGSU
Magnitude range: 7 to 15.5
Period: 11,900d
RA: 20.07.36   Dec: 17.42.14

The dwarf nova WZ Sagittae is one of the most interesting cataclysmic dwarf novae that can be observed by amateurs. The binary consists of a white dwarf primary being orbited by a low mass companion that is an L2 class object indicating that it is a brown dwarf! Studies by the Hubble Space Telescope indicate that the white dwarf primary is about 0.85 times the Sun's mass, while the companion is only 0.08 solar masses.

The system is classed as an SU Ursae Majoris-type star after the progenitor of its class, but the presence of a brown dwarf makes the system much more interesting for observers. The star system is generally very faint and requires a large telescope at minima to see it at all. Observers of WZ Sagittae report superhumps in the light curve that reveal the presence of a precessing accretion disc around the white dwarf. The two objects are so close to each other that the orbital period has been calculated at just 1.3 hours!

Careful observation of the field is needed here, as there are so many Milky Way stars that even the finder charts can be confusing!

WZ Sagittæ

200317   20' FIELD

(1950) 20ʰ 05ᵐ +17° 33'

	Mag.		Mag.		Mag.
A	5·26	K	8·46	U	12·70
B	5·32	L	8·78	W	12·89
C	5·87	M	9·12	X	13·26
D	6·17	N	9·41	Y	13·57
E	6·94	P	9·97	Z	14·27
F	7·31	R	10·34	AA	14·51
G	7·69	S	10·93	BB	14·87
H	7.85	T	11·80		

Based on P.E. Sequence by Webbink

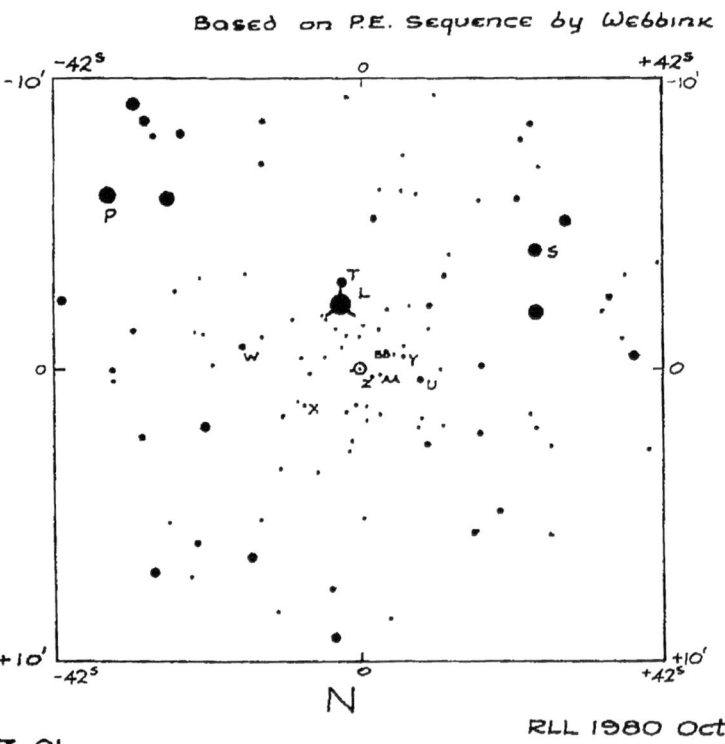

RLL 1980 Oct.

023.01

Fig. 14.45 WZ Sagittae

(Finder chart courtesy of the British Astronomical Association Variable Star Section)

For some additional variable stars for observation or research, try the following recommendations from the BAA.

Star	Type	Range	Period	Frequency
W Cyg	SRb	5.0–7.6	131d	5d
CH Cyg	ZAND+SR	5.6–10.5	NA	Nightly
ST Her	SRb	7.0–8.7	148d	7d
V Vul	RVA	8.1–9.5	76d	5d-7d

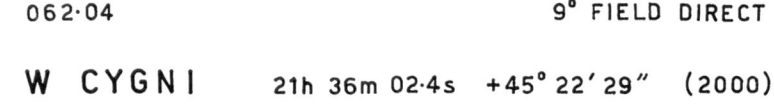

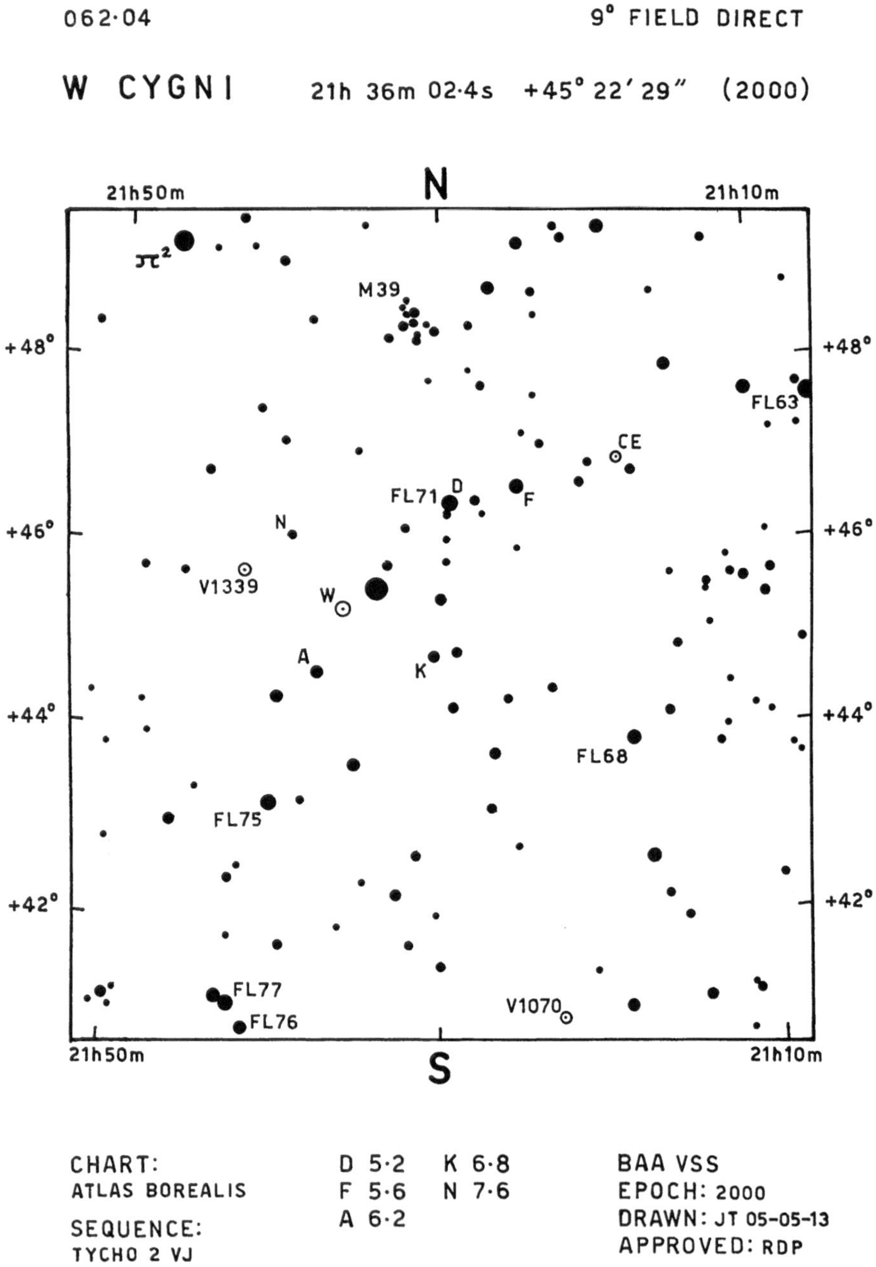
062·04
9° FIELD DIRECT
W CYGNI 21h 36m 02·4s +45° 22′ 29″ (2000)
N
21h50m
21h10m
+48°
+46°
+44°
+42°
M39
FL63
CE
FL71
D
F
N
V1339
W
A
K
FL68
FL75
FL77
FL76
V1070
S
CHART:
ATLAS BOREALIS
SEQUENCE:
TYCHO 2 VJ
D 5·2 K 6·8
F 5·6 N 7·6
A 6·2
BAA VSS
EPOCH: 2000
DRAWN: JT 05-05-13
APPROVED: RDP

Fig. 14.46 W Cygni

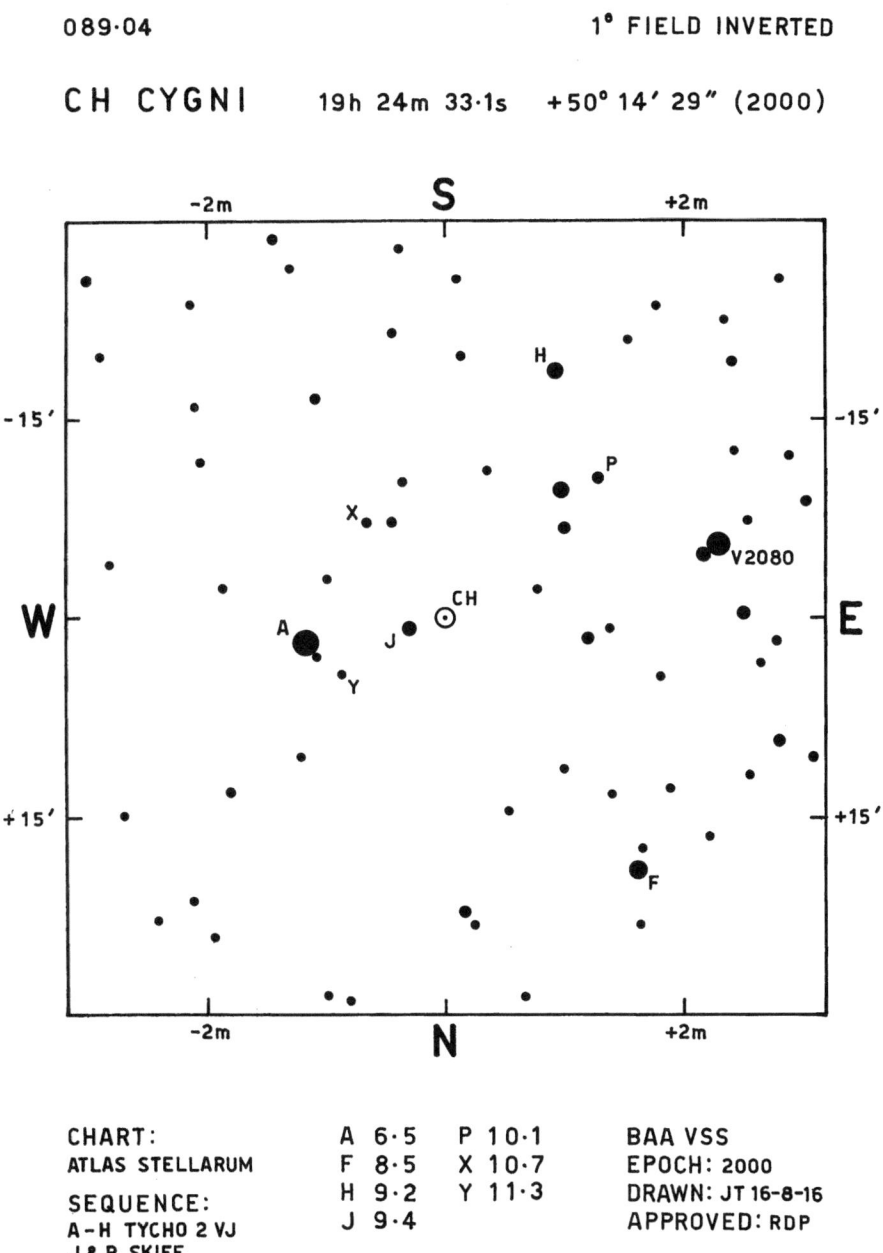

089·04   1° FIELD INVERTED

CH CYGNI   19h 24m 33·1s   +50° 14′ 29″ (2000)

CHART:	A 6·5	P 10·1	BAA VSS
ATLAS STELLARUM	F 8·5	X 10·7	EPOCH: 2000
	H 9·2	Y 11·3	DRAWN: JT 16-8-16
SEQUENCE:	J 9·4		APPROVED: RDP
A-H TYCHO 2 VJ			
J & P SKIFF			
X & Y SRO			

Fig. 14.47 CH Cygni

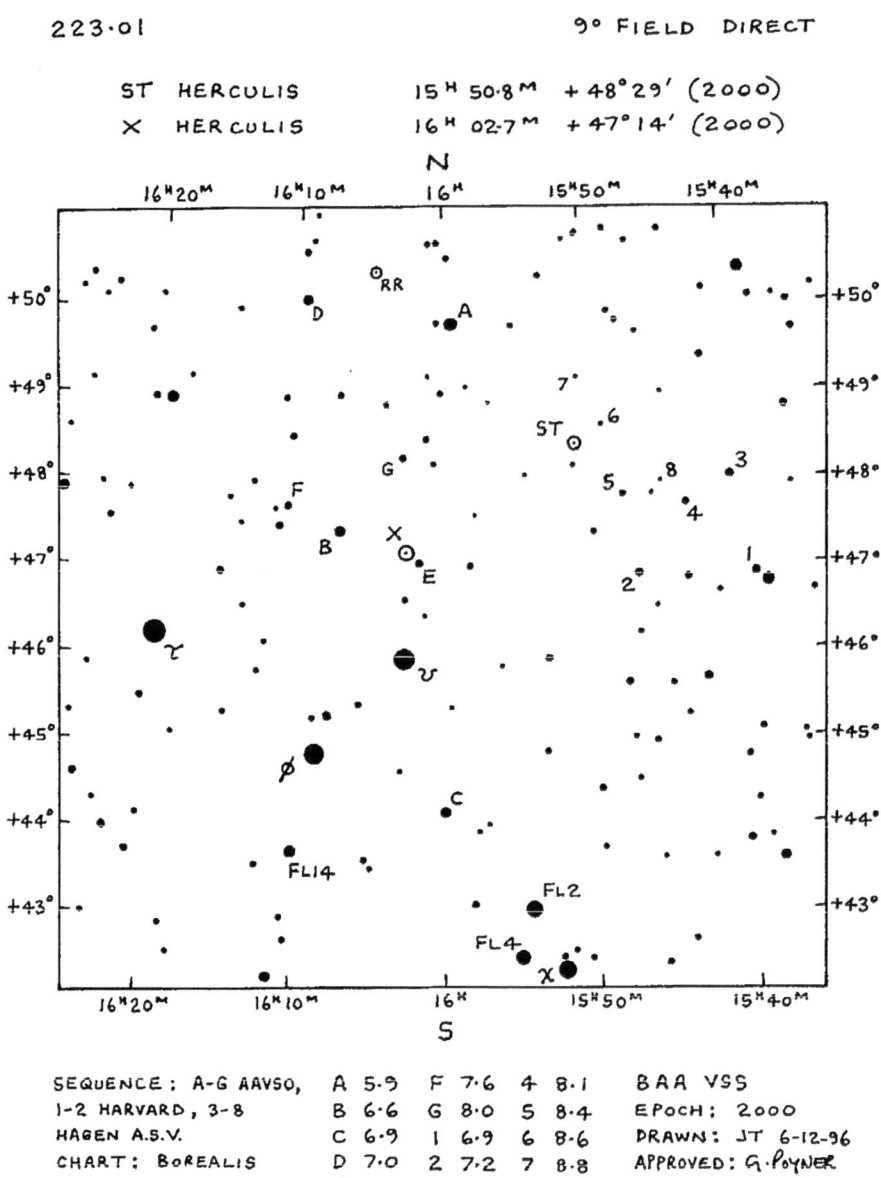

223·01   9° FIELD   DIRECT

ST HERCULIS   15ᴴ 50·8ᴹ + 48° 29' (2000)
X   HERCULIS   16ᴴ 02·7ᴹ + 47° 14' (2000)

SEQUENCE : A-G AAVSO,   A 5·9   F 7·6   4 8·1   BAA VSS
1-2 HARVARD , 3-8   B 6·6   G 8·0   5 8·4   EPOCH : 2000
HAGEN A.S.V.   C 6·9   1 6·9   6 8·6   DRAWN : JT 6-12-96
CHART : BOREALIS   D 7·0   2 7·2   7 8·8   APPROVED : G.Poyner
                         E 7·4   3 7·7   8 9·0

Fig. 14.48 ST Herculis

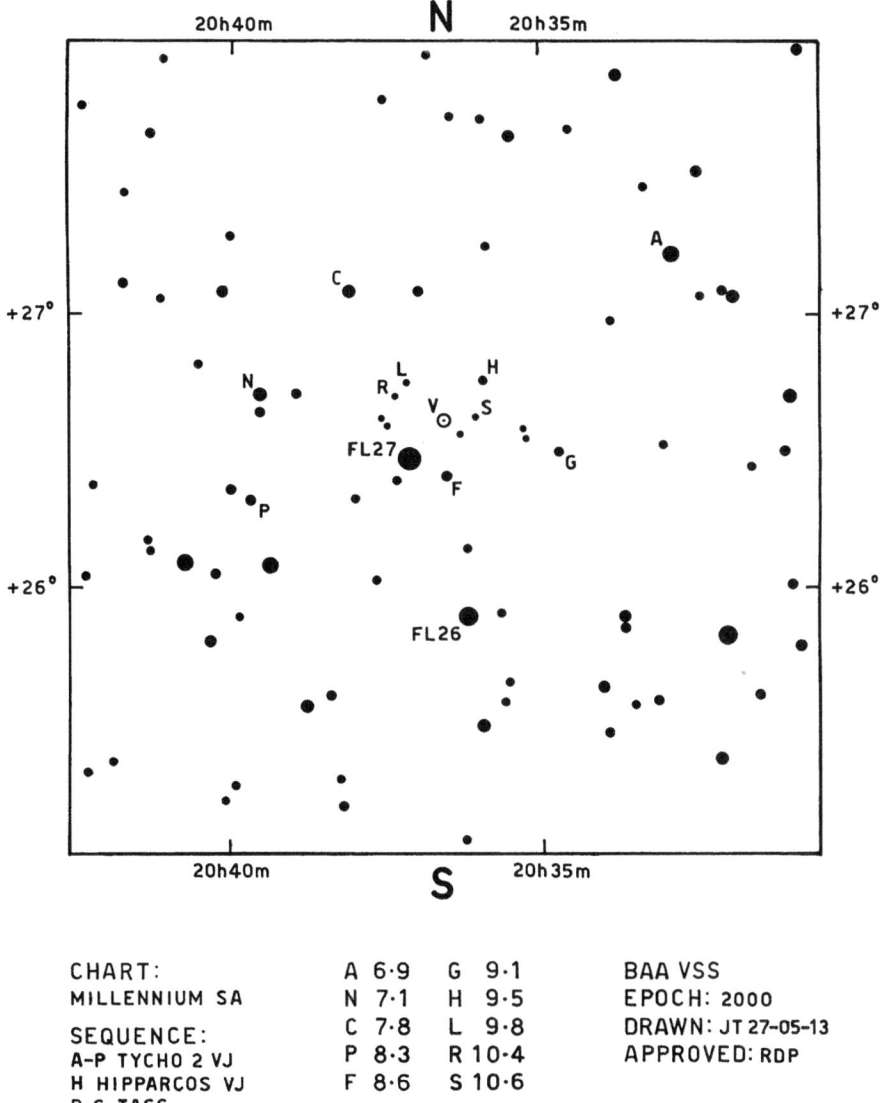

058·02   3° FIELD DIRECT

V VULPECULAE   20h 36m 32·0s +26° 36′ 14″ (2000)

CHART:		A 6·9	G 9·1	BAA VSS
MILLENNIUM SA		N 7·1	H 9·5	EPOCH: 2000
SEQUENCE:		C 7·8	L 9·8	DRAWN: JT 27–05–13
A-P TYCHO 2 VJ		P 8·3	R 10·4	APPROVED: RDP
H HIPPARCOS VJ		F 8·6	S 10·6	
R-S TASS				

Fig. 14.49 V Vulpecula

(Finder charts courtesy of the British Astronomical Association Variable Star Section)

GAMMA CASSIOPEIAE
Type: GCAS
Magnitude range: 1.6 to 3
Period: –
RA: 00.56.42   Dec: 60.43.00

Gamma Cassiopeiae is an eruptive-type variable star whose apparent magnitude changes irregularly between +1.6 and +3.0. It is the prototype of the class of Gamma Cassiopeiae variable stars, and in the late 1930s it underwent an expulsion of a shell of gas that altered its spectral characteristics. Its brightness increased to above magnitude 2.0, before dropping back to magnitude 3.4. Since then, the star has been gradually brightening and now shines on average at magnitude 2.2. However, during its peak luminosity, gamma Cassiopeia outshines both alpha and beta Cassiopeiae at magnitude 2.2 and 2.3, respectively

The variable is a rapidly spinning star, giving it a pronounced equatorial bulge, resulting in it losing materials in expanding shells that come off the bulge. The ejected matter then forms a hot circumstellar disc of gas that makes the light output vary.

The spectrum of this massive star is given as B0.5 IVe. This is a subgiant star that has reached a stage of its evolution where it is exhausting the supply of hydrogen in its core region and transforming into a giant star. The "e" suffix is used for stars that show emission lines of hydrogen in the spectrum, caused in this case by the circumstellar disc. According to the catalogues, Gamma Cassiopeia has 17 times the Sun's mass and is radiating as much energy as 34,000 suns. Its outer atmosphere has a temperature of 25,000 K, and its period of variability is irregular and dependent on the shell ejection.

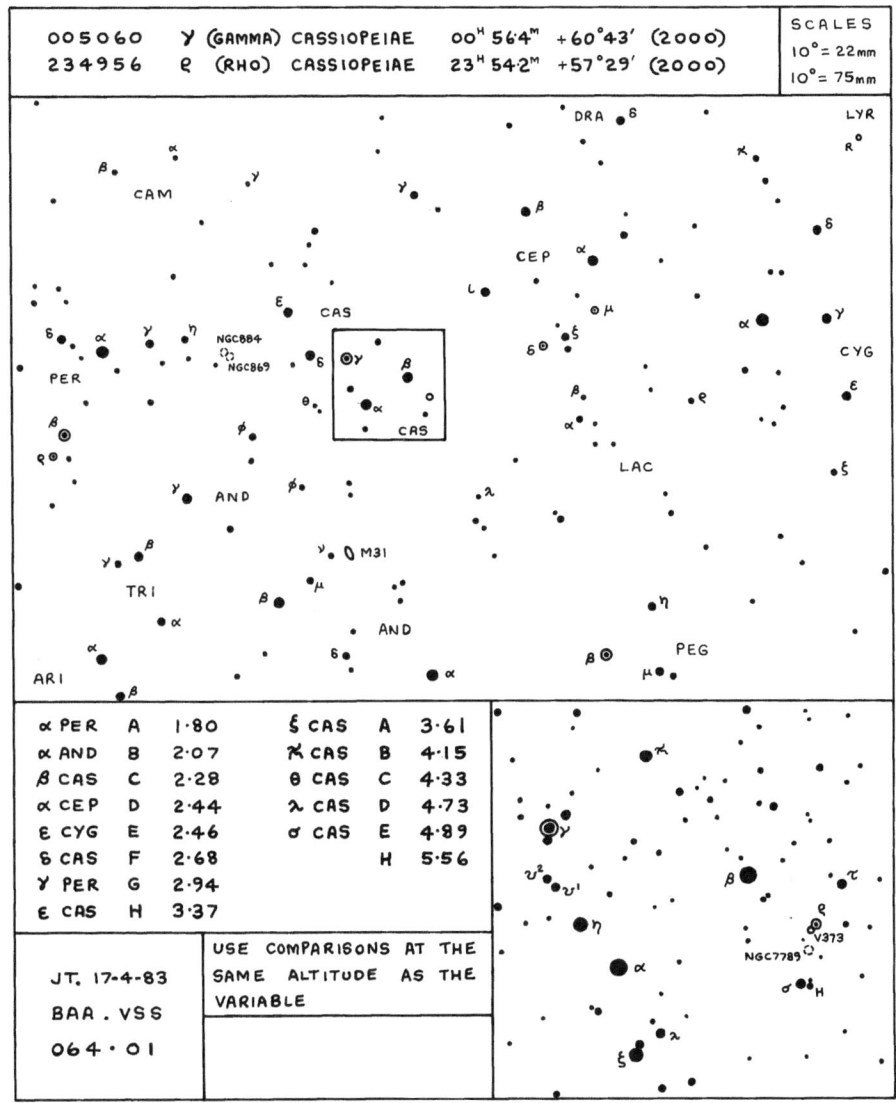

Fig. 14.50 Gamma Cassiopeiae

(Finder chart courtesy of the British Astronomical Association Variable Star Section)

MU CEPHEI
Type: SRC
Magnitude range: 3.4 to 5.1
Period: –
RA: 21.43.30   Dec: 58.46.48

William Herschel's famous "Garnet Star" is a huge red giant in the constellation Cepheus that is visible to the naked eye and irregularly varies in light output and in color. Its usual hue is an orange-red, but this intensifies at minimum light to deep red. Mu Cephei is one of the largest stars known and must be very luminous, as its estimated distance is close to 10,000 light years away, and it has an estimated radius of 1,260 times that of the Sun.

Mu Cephei is a semi-regular variable and is fusing helium into carbon internally so it is approaching the end of its life. However, it is expected to lose its outer envelope and become a luminous blue variable before possibly exploding as a supernova. Given that its estimated mass is 20 times that of the Sun, unless it loses a substantial part of its mass, Mu Cephei will become a stellar black hole.

In long-exposure photographs and in infrared, there is a surrounding torus of gas that extends 6 arcseconds away from the star, indicating mass loss over the last few thousand years.

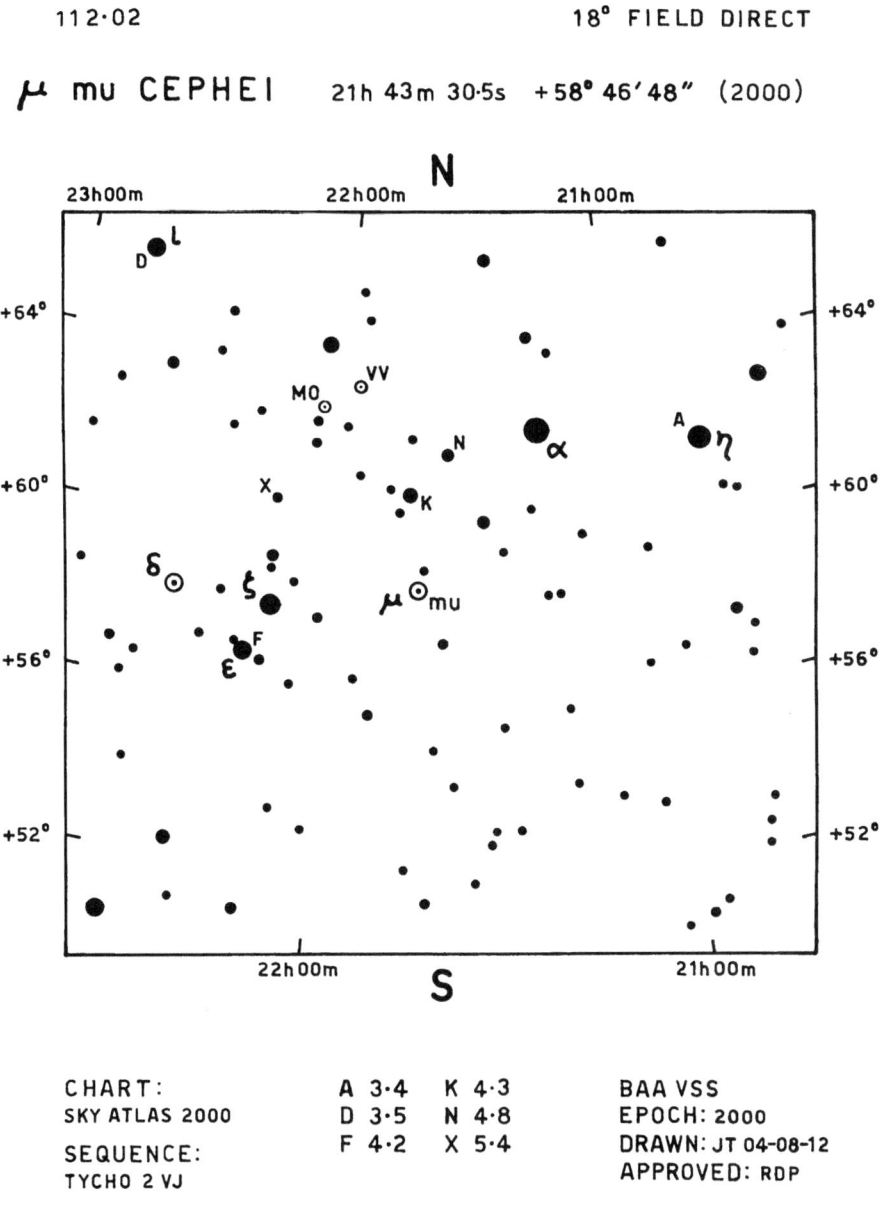

112·02   18° FIELD DIRECT

μ mu CEPHEI   21h 43m 30·5s +58° 46′ 48″ (2000)

CHART:	A 3·4	K 4·3	BAA VSS
SKY ATLAS 2000	D 3·5	N 4·8	EPOCH: 2000
	F 4·2	X 5·4	DRAWN: JT 04-08-12
SEQUENCE:			APPROVED: RDP
TYCHO 2 VJ			

Fig. 14.51 Mu Cephei

(Finder chart courtesy of the British Astronomical Association Variable Star Section)

RHO CASSIOPEIAE
Type: SRD
Magnitude range: 4.1 to 6.2
Period: 320d
RA: 23.54.23   Dec: 57.29.58

Rho Cassiopeia is one of the largest stars we can see with the naked eye and is an extremely luminous object, shining with at least 500,000 times the Sun's output. Lying 8,200 light years away this yellow hypergiant star is one of only a dozen known and was thought to irregularly dim between magnitudes 4.3 and 6. In the year 2000–2001, the star was shown to have been subject to eruptions that gave rise to variability.

The eruptions result from the vast size of the star, being very close to the Eddington limit, in which the radiation from the star disrupts the gas that gravity just cannot hold onto. Yellow hypergiants such as Rho Cass lie in a temperature range where opacity variations in zones of partial ionization of hydrogen and helium cause pulsations. Similar to those of Cepheid variable stars in hypergiants, these pulsations are generally irregular and small, but combined with the overall instability of the outer layers of the star they can possibly result in larger outbursts.

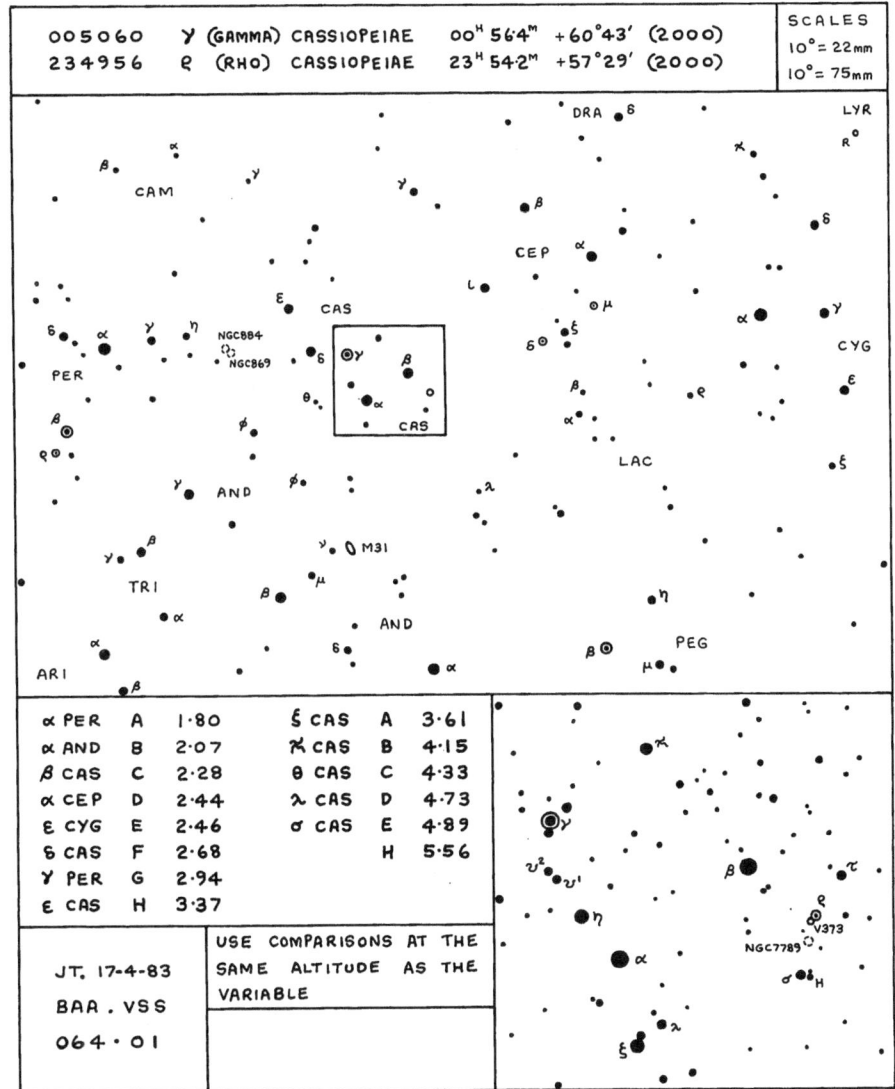

Fig. 14.52 Rho Cassiopeia

(Finder chart courtesy of the British Astronomical Association Variable Star Section)

RY URSAE MAJORIS
Type: SRB
Magnitude range: 6.7 to 8.3
Period: 310d
RA: 12.18.04   Dec: 61.35.20

Another semi-regular Mira-type variable star that for observers in higher northern latitudes never sets. Typical of the Mira types is the red color and the rather indistinct maximum and minimum, which do not reach the heights or depths that are always described in books and journals. Occasionally, RY Uma has been seen with the naked eye at maximum, but as the range is not huge, the whole light curve can be plotted with a pair of binoculars if the observer is experienced.

217·02   9° FIELD DIRECT

RY URSAE MAJORIS 12h 20m 27·4s +61° 18' 35" (2000)
Z URSAE MAJORIS 11h 56m 30·2s +57° 52' 18" (2000)

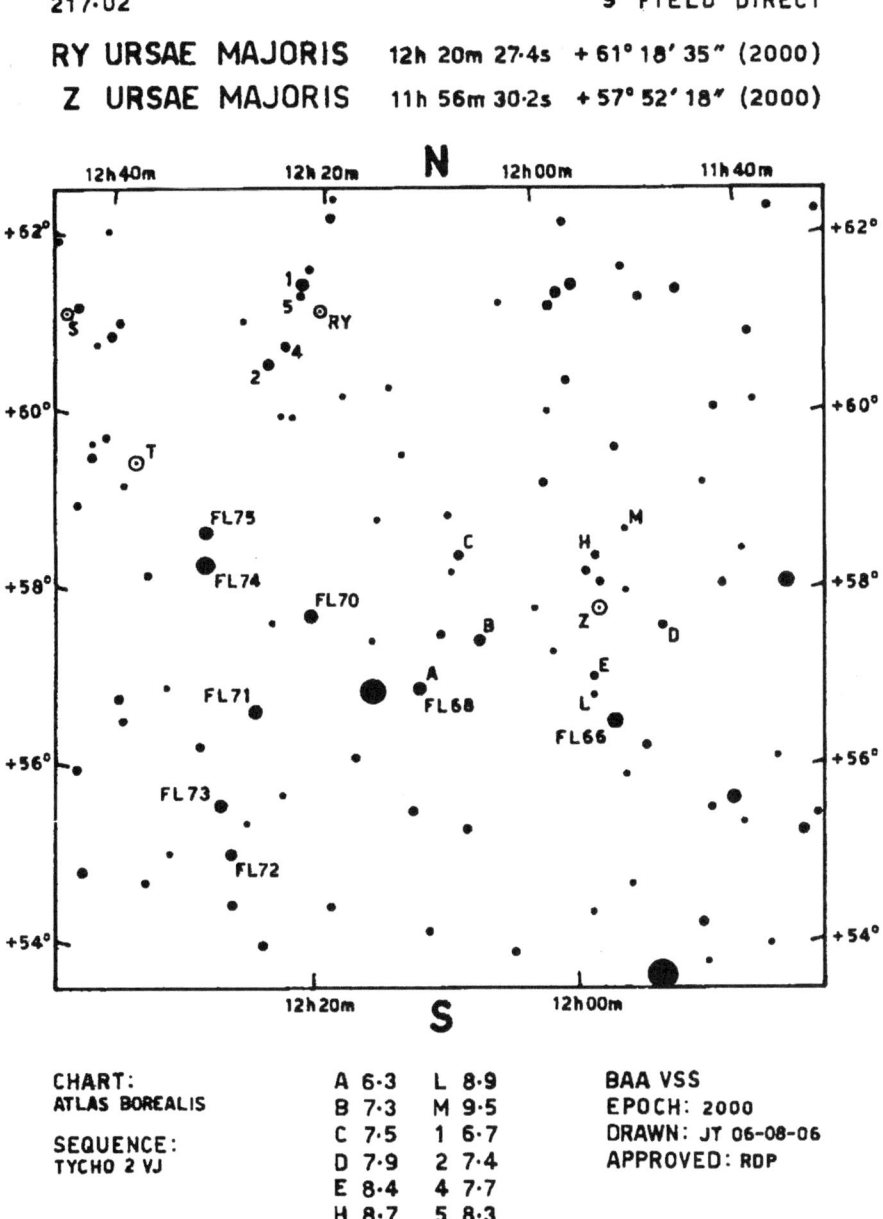

CHART:	A 6·3	L 8·9	BAA VSS
ATLAS BOREALIS	B 7·3	M 9·5	EPOCH: 2000
	C 7·5	1 6·7	DRAWN: JT 06-08-06
SEQUENCE:	D 7·9	2 7·4	APPROVED: RDP
TYCHO 2 VJ	E 8·4	4 7·7	
	H 8·7	5 8·3	

Fig. 14.53 RY Ursae Majoris

(Finder chart courtesy of the British Astronomical Association Variable Star Section)

RZ CASSIOPEIAE
Type: EA/SD
Magnitude range: 6.2 to 7.7
Period: 1.19d
RA: 02.44.23   Dec: 69.25.30

RZ Cassiopeia is an eclipsing binary star that can be seen in binoculars with a range of just about 1.5 magnitudes between maximum and minima. The star has been studied by astronomers for many years, and between the stars there is a complex interchange of gas and some ejection from the system to form complex shells that show up in absorption and emission spectra. The decline to minimum light is fairly quick, and observers are prompted to look at the star and estimate its brightness every 15 minutes, when an eclipse is predicted. Times of eclipse are available from the BAA and the AAVSO.

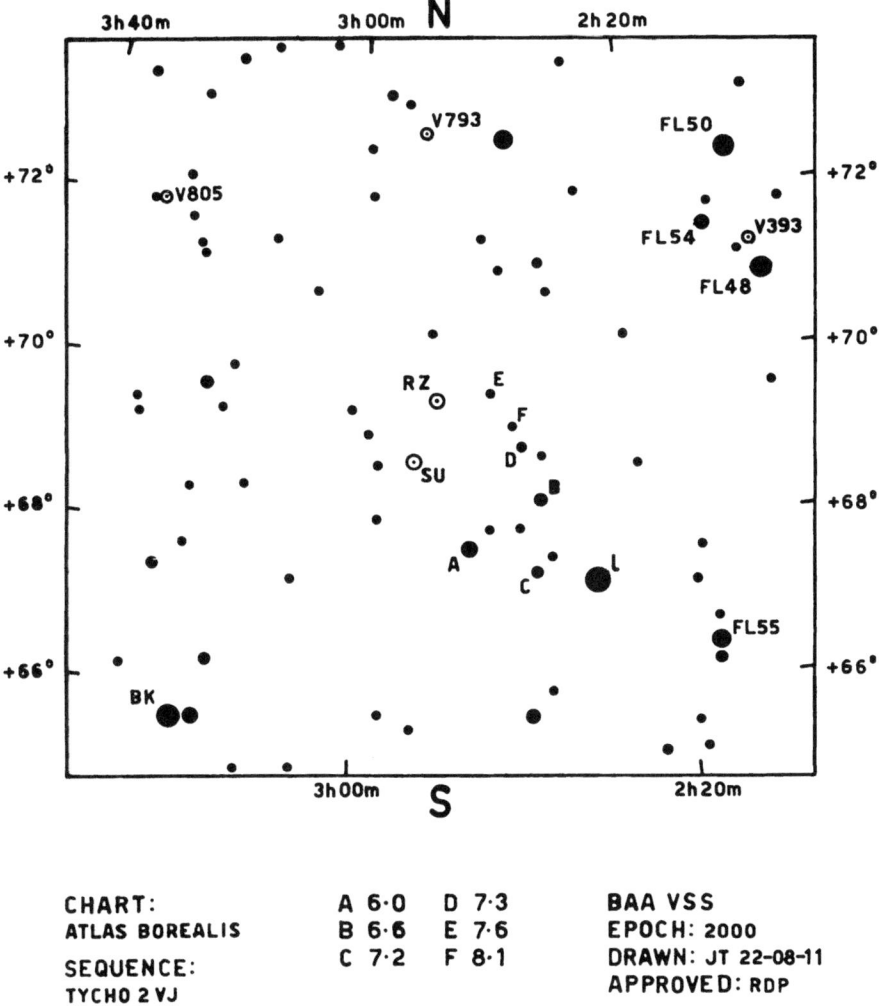

236·02

9° FIELD DIRECT

RZ CASSIOPEIAE 02h 48m 55·5s +69° 38′03″ (2000)

CHART: A 6·0   D 7·3   BAA VSS
ATLAS BOREALIS   B 6·6   E 7·6   EPOCH: 2000
SEQUENCE: C 7·2   F 8·1   DRAWN: JT 22–08–11
TYCHO 2 VJ   APPROVED: RDP

Fig. 14.54 RZ Cassiopeiae

(Finder chart courtesy of the British Astronomical Association Variable Star Section)

Z URSAE MAJORIS
Type: SRB
Magnitude range: 6.2 to 9.4
Period: 196d
RA: 11.53.54   Dec: 58.09.02

Z Ursae Majoris is a red giant star whose semi-regular brightness variations are due to pulsations in its outer layers. Z UMa can be relied upon to produce a good amount of brightness variation on a regular basis and has two distinct periods that relate to the harmonic modes of the star's outer layers, revealing a period of 98 days and 196 days with alternating and overlapping deep and shallow minima. The large range of Z Uma makes it a favorite target for binocular observers, and those with a small telescope can follow its variations easily and build their confidence in variable star observing if they are enjoying it for the first time.

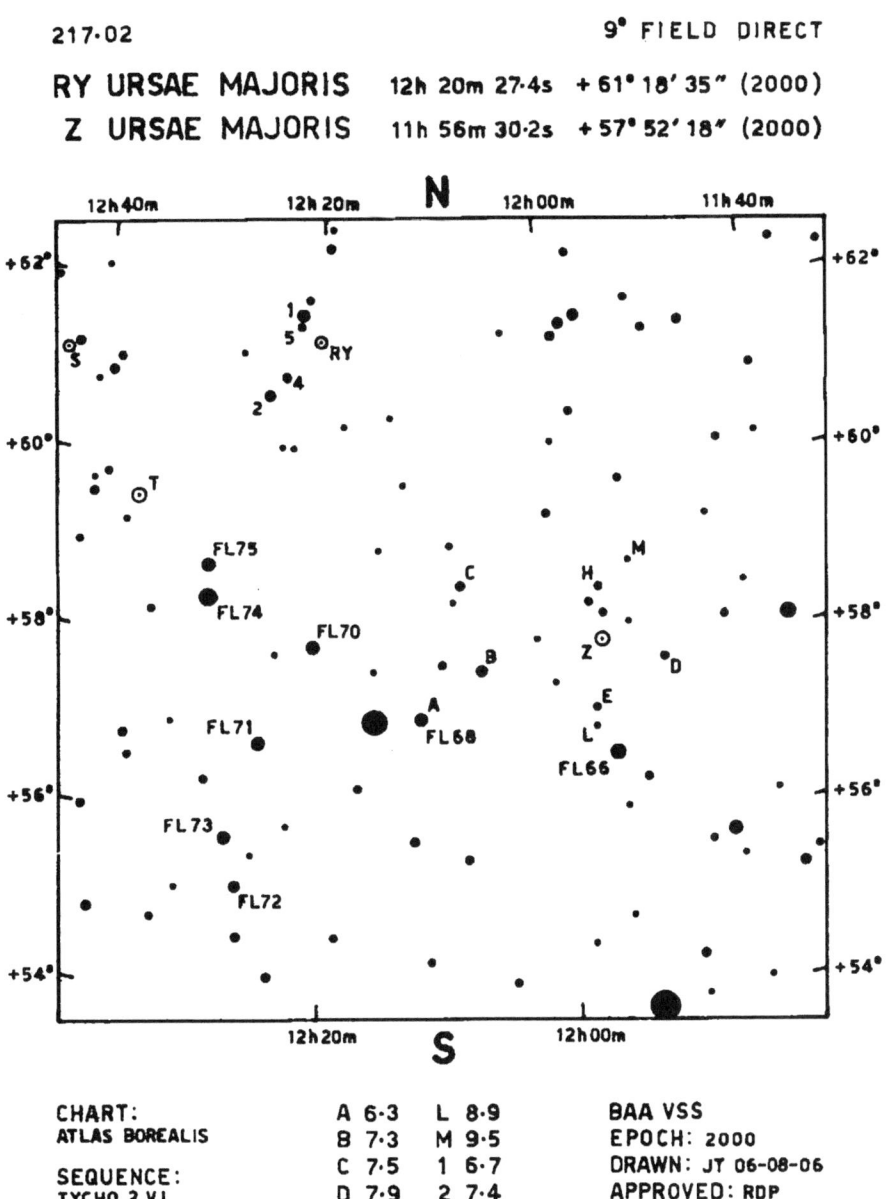

217·02   9° FIELD DIRECT

RY URSAE MAJORIS   12h 20m 27·4s   + 61° 18′ 35″ (2000)
Z URSAE MAJORIS   11h 56m 30·2s   + 57° 52′ 18″ (2000)

CHART:	A 6·3	L 8·9	BAA VSS
ATLAS BOREALIS	B 7·3	M 9·5	EPOCH: 2000
	C 7·5	1 6·7	DRAWN: JT 06-08-06
SEQUENCE:	D 7·9	2 7·4	APPROVED: RDP
TYCHO 2 VJ	E 8·4	4 7·7	
	H 8·7	5 8·3	

Fig. 14.55 Z Ursae Majoris

(Finder chart courtesy of the British Astronomical Association Variable Star Section)

For some additional variable stars for research, why not try the following in this table?

Star	Type	Range	Period	Frequency
T Cas	Mira	6.9–13.0	445d	10d-14d
TX Dra	SRb	6.6–8.4	78d?	5d
RY Dra	SRb?	6.0–8.2	200d	7d
U Cam	SRb	7.7–8.8	Unknown	5d

067·02   2° FIELD INVERTED

T CASSIOPEIAE   00h 23m 14·3s   +55°47′33″ (2000)

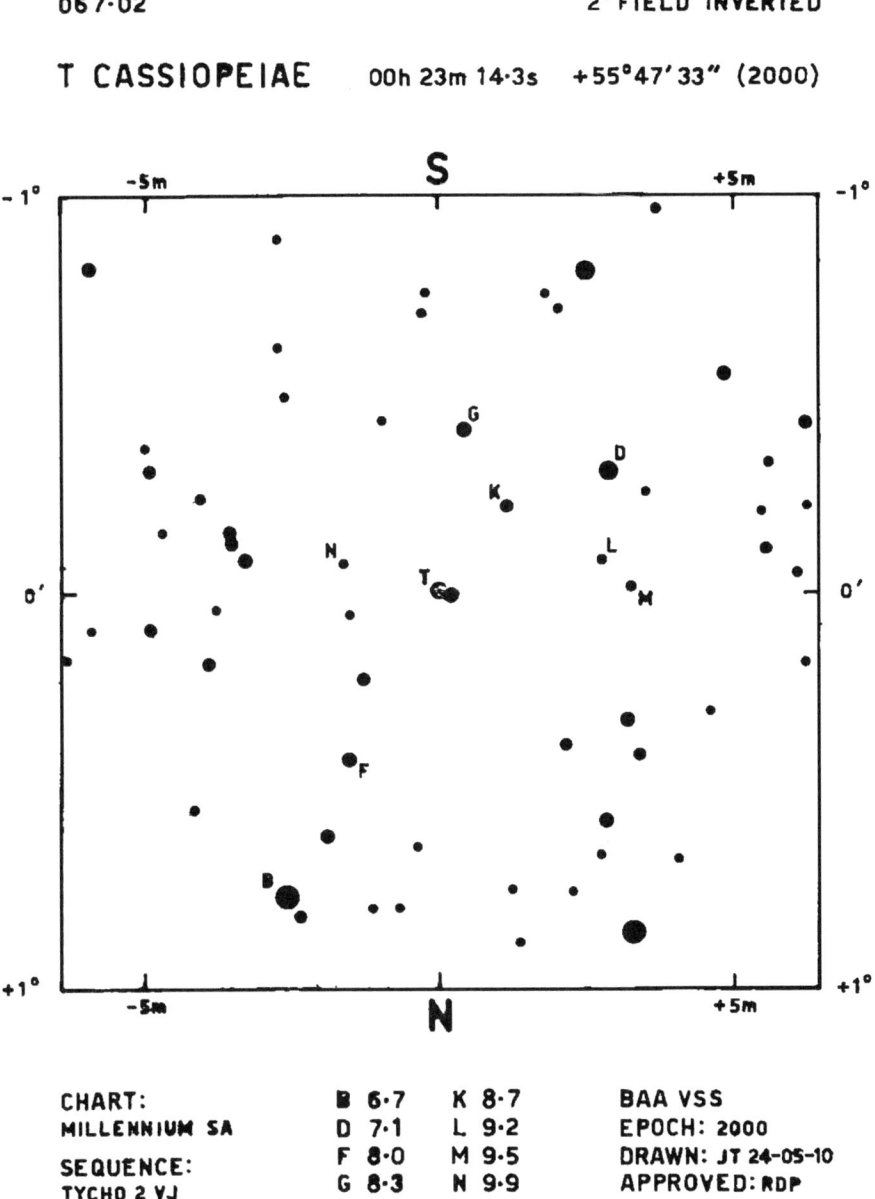

CHART: B 6·7   K 8·7   BAA VSS
MILLENNIUM SA   D 7·1   L 9·2   EPOCH: 2000
SEQUENCE: F 8·0   M 9·5   DRAWN: JT 24-05-10
TYCHO 2 VJ   G 8·3   N 9·9   APPROVED: RDP

Fig. 14.56 T Cassiopeia

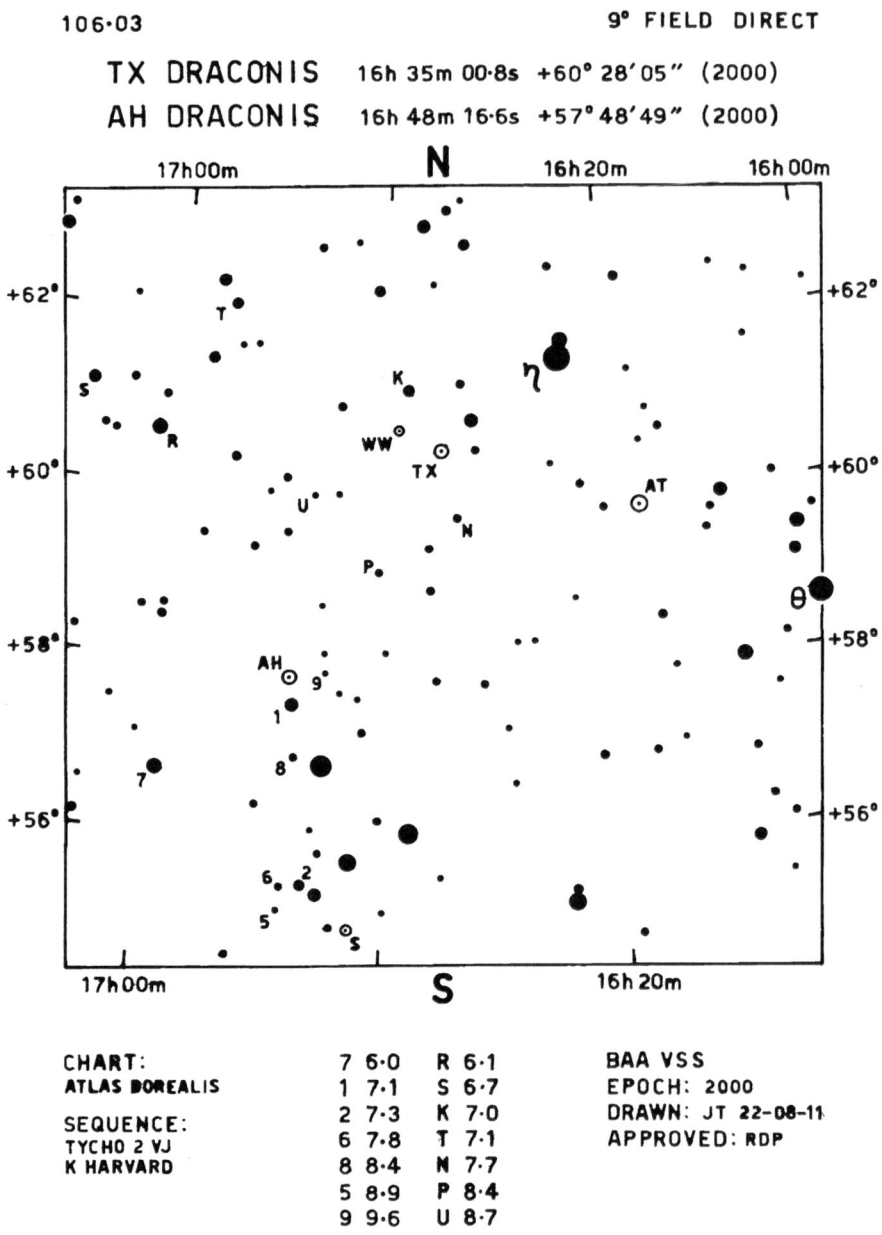

9° FIELD DIRECT

TX DRACONIS   16h 35m 00·8s +60° 28′05″ (2000)
AH DRACONIS   16h 48m 16·6s +57° 48′49″ (2000)

CHART:		7 6·0	R 6·1	BAA VSS
ATLAS BOREALIS		1 7·1	S 6·7	EPOCH: 2000
		2 7·3	K 7·0	DRAWN: JT 22-08-11
SEQUENCE:		6 7·8	T 7·1	APPROVED: RDP
TYCHO 2 VJ		8 8·4	N 7·7	
K HARVARD		5 8·9	P 8·4	
		9 9·6	U 8·7	

Fig. 14.57 TX Draconis

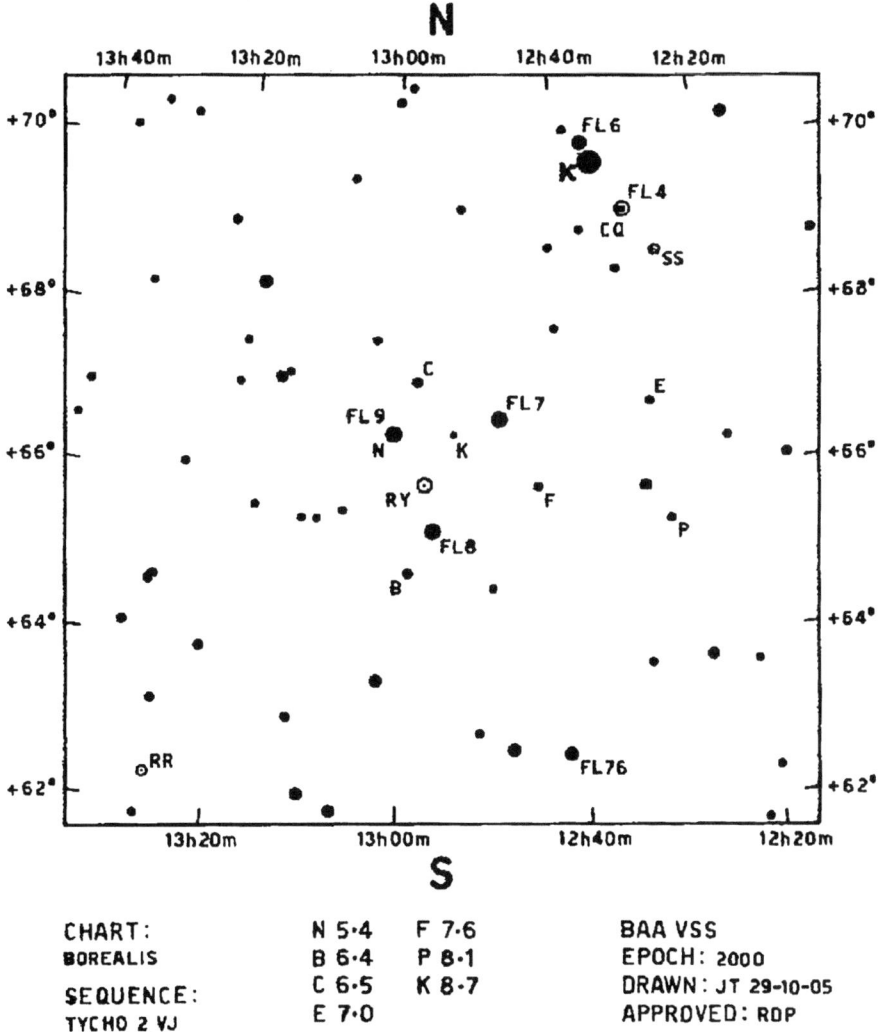

Fig. 14.58 RY Draconis

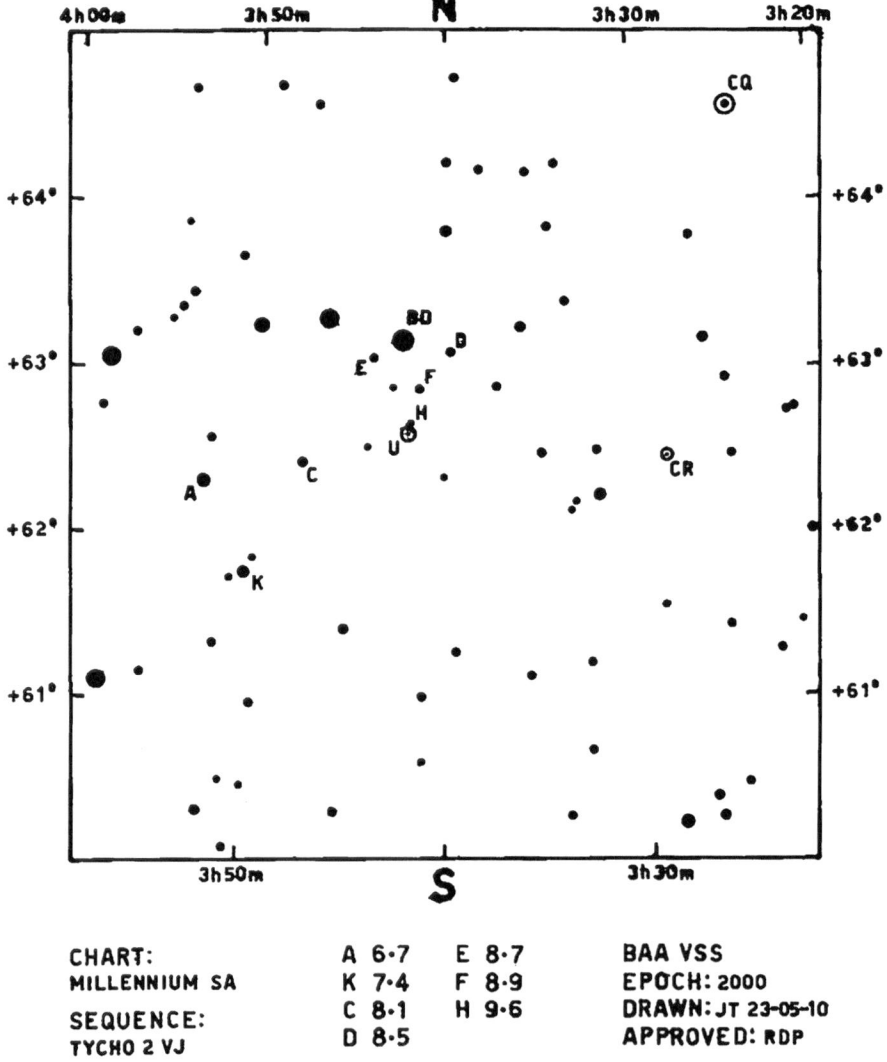

100·02   5° FIELD DIRECT

U CAMELOPARDALIS   03h 41m 48·2s +62°38′54″ (2000)

CHART:	A 6·7	E 8·7	BAA VSS
MILLENNIUM SA	K 7·4	F 8·9	EPOCH: 2000
SEQUENCE:	C 8·1	H 9·6	DRAWN: JT 23-05-10
TYCHO 2 VJ	D 8·5		APPROVED: RDP

Fig. 14.59 U Camelopardalis

(Finder charts courtesy of the British Astronomical Association Variable Star Section)

Not included here in this list are the following stars in this quick list for observation. The individual stars can be researched by the observer from such sources as the AAVSO or the BAA and volumes such as *Burnham's Celestial Handbook* fill in some of the details on such stars as to their evolution and types.

Quick List of Additional Variable Stars for Observation

Star	Type	Magnitude	Period (d)
SS Aur	UG	10.3–15.8	55.5
S Car	M	4.5–9.9	149
T Cas	M	6.9–13.0	445
V Cas	M	6.9–13.4	229
T Cen	SR	5.5–9.0	90
R Cen	M	5.3–11.8	546
S CMi	M	6.6–13.2	333
R CrB	RCB	5.7–14.8	N/A
chi Cyg	M	3.3–14.2	408
R Cyg	M	6.1–14.4	426
AB Dra	UG	11.0–15.3	13.4
T Her	M	6.8–13.7	165
W Her	M	7.6–14.4	280
R Hor	M	4.7–14.3	408
T Hor	M	7.2–13.7	218
R LMi	M	6.3–13.2	372
R Leo	M	4.4–11.3	310
R Oph	M	7.0–13.8	306
R Pic	SR	6.3–10.1	171
R Ser	M	5.2–14.4	356
R Sgr	M	6.7–12.8	269
RY Sgr	RCB	5.8–14.0	N/A
T Tuc	M	7.5–13.8	250
S UMa	M	7.1–12.7	226
T UMa	M	6.6–13.5	257
R Vul	M	7.0–14.3	137

Appendix:
Spectroscopy
of Variable Stars

As amateur astronomers gain ever-increasing access to professional tools, the science of spectroscopy of variable stars is now within reach of the experienced variable star observer. In this section we shall examine the basic tools used to perform spectroscopy and how to use the data collected in ways that augment our understanding of variable stars. Naturally, this section cannot cover every aspect of this vast subject, and we will concentrate just on the basics of this field so that the observer can come to grips with it.

It will be noticed by experienced observers that variable stars often alter their spectral characteristics as they vary in light output. Cepheid variable stars can change from G types to F types during their periods of oscillation, and young variables can change from A to B types or vice versa. Spectroscopy enables observers to monitor these changes if their instrumentation is sensitive enough.

However, this is not an easy field of study. It requires patience and dedication and access to resources that most amateurs do not possess. Nevertheless, it is an emerging field, and should the reader wish to get involved with this type of observation know that there are some excellent guides to variable star spectroscopy via the BAA and the AAVSO. Some of the workshops run by Robin Leadbeater of the BAA Variable Star section and others such as Christian Buil are a very good introduction to the field.

© Springer Nature Switzerland AG 2018
M. Griffiths, *Observer's Guide to Variable Stars*, The Patrick Moore
Practical Astronomy Series, https://doi.org/10.1007/978-3-030-00904-5

Spectra, Spectroscopes and Image Acquisition

What are spectra, and how are they observed? The spectra we see from stars is the result of the complete output in visible light of the star (in simple terms). This output produces a variety of colors, from blue at the short wavelength end of the spectrum to red at the longer wavelengths of the visible spectrum. To obtain the spectra of stars we need either a prism that will bend or refract the light into its composite colors or a diffraction grating, which is an instrument with hundreds of fine lines or grooves that split the light into its composite wavelengths.

Most spectrographs are equipped with a diffraction grating. Some can be obtained cheaply, or commonly available filters can be obtained and then readjusted to perform the same function as a spectroscope.

Alongside the usual equipment of CCD camera, telescopes, driven mounts and the software to control them all will be the consideration of what spectroscope to use. There are no spectroscopes in general use that will suffice for this exercise, so some observers use a Cokin filter (number 40), which has approximately 250 grooves per millimeter, that can act as a diffraction grating in front of your CCD camera. It is useful to remember that such equipment may not pick up the fainter stars, and so spectroscopy of brighter stars should be attempted first as part of the observers' learning curve.

Observers who perform spectroscopy have a variety of tools at their disposal. Many use nothing more than a relatively inexpensive spectrographic tool known as a Star Analyzer, which has a resolution of about 500 nm and can be fitted to the CCD in the same way as the Cokin filter above. However, most spectroscopes are expensive pieces of equipment, and some for laboratory spectroscopy can be modified for astronomical use. Such items as the Elliot Institute CCD Spec, the Shelyak LHIRES and the JTW L200 are quite pricey but have higher resolutions down to 5 nm. Some of these are shown in Fig. A.1.

Depending on your preference, any spectroscope will smear the images from a light source, and the resolution of the image will depend on the FWHM of the images, the brightness of your subject variable star and the optical train of your telescope and camera. Observers in this field often add a small slit to the front of the grating or CCD so that the image is not overlaid with alternate stars, and the resolution is a little more constant per field. As a general rule of thumb, the higher the dispersion of the spectrum (the spread of light) the brighter the target variable star must be to get good information and data.

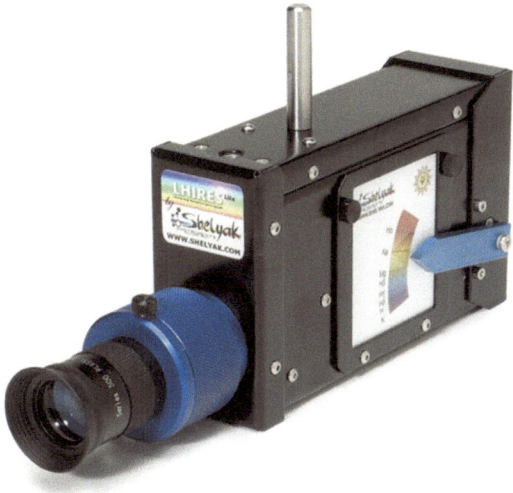

Fig. A.1 LHIRES spectroscope. (Image from http://www.astronomy.com/observing/product-reviews/2009/12/lhires-lite-spectroscope.)

Software used by many observers include the ISIS image subtraction package for photometry and comparison with known spectra via the MILES software library, which can be accessed at http://www.iac.es/proyecto/miles/pages/stellar-libraries/miles-library.php. There are some excellent guides to exploring and using this software available at http://www.astrosurf.com/buil/isis/guide_lhires/tuto1_en.htm.

All files are in FITS format, so that immediate comparison can be made with your own spectrograph CCD files. ISIS can not only align the multiple CCD exposures you may have taken of your variable star but can also remove the sky background and reduce the data to a curve that enables the observer to see the most prevalent lines in the spectrum of the target star, as one can see in Fig. A.2 here. Using Wein's law can then enable you to discover what type of star you are examining (although of course you will know this already!) and to confirm the spectral type for your own records.

Although doing spectroscopy at this level seems like re-inventing the wheel, it is an important adjunct to amateur research, and we must remember that many variable stars are not followed by professional astronomers. Any information gleaned from such variable stars via spectroscopy can only add to the growing amount of knowledge that we have of the cosmos. Although the spectroscopy being done here is relatively low resolution, one never knows when such data may be useful in the future. Additionally, supernova research can gain a lot from the approach of amateurs using spectroscopy,

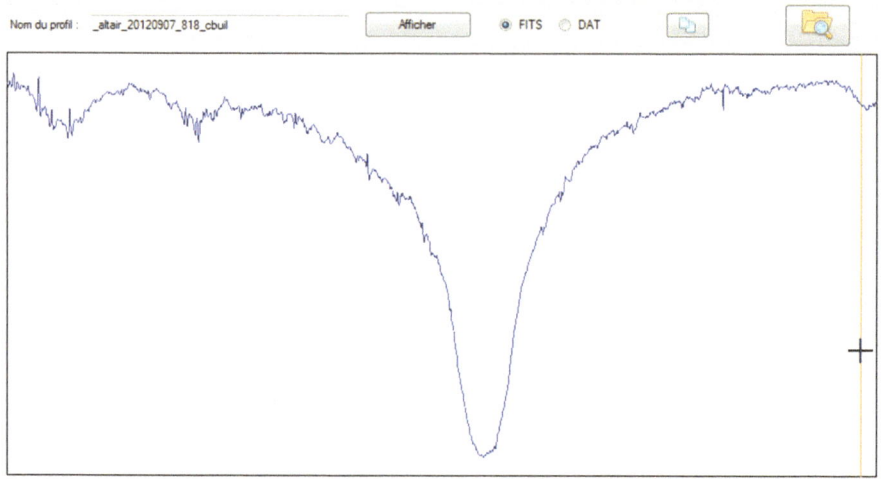

Fig. A.2 ISIS spectrum. (Image from http://www.astrosurf.com/buil/isis/eshel/reduction/echelle.htm.)

and many of these observers are combing the skies looking for such events. In fact, it is mostly amateurs that spot supernovae, as they have the time, abilities and resources necessary to perform this function. Spectra of a supernova occurring is important in ascertaining the type of object it is (Type I or II); recognition of this will be a feather in the cap of any supernova patroller.

Some basics in our approach to spectroscopy of variable stars follow the observational techniques provided earlier in the book. It is important in amateur spectroscopy to take the image when the object under study is at culmination, thus reducing air mass and ensuring maximum image quality. If using a CCD, make sure that the chip does not become saturated. As in all CCD imaging, flats and darks can be taken and subtracted. Binning is also useful to ensure that you have a reasonable pixel count. Additionally, one is still dealing with an image that requires as much input as possible in much the same way as photometry, so the FWHM of the source should be accounted for.

Once you have the field of view of your subject star, your image may look a little like Fig. A.3, with stars smeared into rainbow-like trails. This is not a mistake. Hidden within the little rainbows (or if shooting in monochrome the extended trail) are all the bits of information you are looking for in the form of absorption bands characterizing the chemical makeup of the star. As this is low resolution, do not expect to get all the

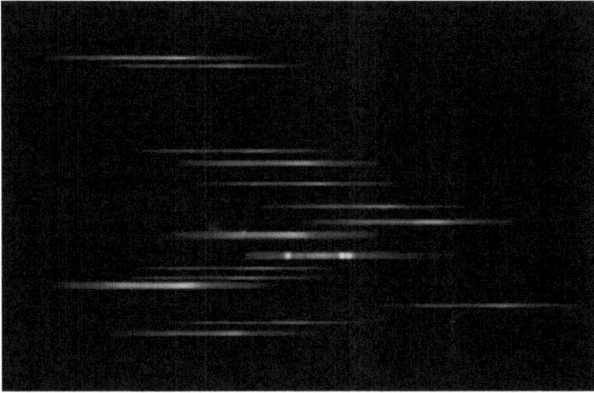

Fig. A.3 Spectra of field of view. (Image from https://www.rspec-astro.com.)

absorption band that could be present, but bands such as H alpha, sodium and oxygen may well be present.

Such details may also be spoiled a little by interstellar dust, making the image redder than it should appear. Remember, too, that hydrogen lines are visible only in the middle spectral ranges, so O- and B-type stars will not have these lines, but you may be able to pick up the lines of helium instead if you have that kind of sensitivity. Different types of variable stars also have special characteristics in their spectra, while dwarf nova and classical nova have very different traits that make them instantly recognizable if one knows something of the astrophysics of the subject.

Variable star spectra should be apparent down to a visual magnitude of 13 or 14 if one is using a CCD and a telescope in the 20-cm range. This enables the observer to gain a large amount of data from several different types of variables, which will be good practice, but you may want to concentrate just on one type of variable star. If you have very good spectra reduced via ISIS or other software then you may wish to follow the next exercise in determining the distance to the source.

Determining Red Shift or Blue Shift

Considering that many variable stars show radial pulsations or have components that move toward or away from the observer, it may be possible with good resolution to make some calculations of the speed of movement of the stellar photosphere or orbiting object. It must be said however that in general, most spectrographs gained from amateur telescopes may not have sufficient resolution to do this exercise, as the changes in blue or red shift toward or away from an observer are going to be fractions of a nanometer at most.

Nonetheless, the speed or velocity of movement can easily be discovered using a basic mathematical relationship. The equation for this relationship is:

$$V_r = c(\lambda' - \lambda)/\lambda$$

where V_r is the radial velocity, λ is the standard laboratory wavelength and λ' is the observed wavelength. C is the velocity of light at 3×10^8 M/s. One would have to have access to standard lab wavelengths in H alpha and other prominent lines in the stellar spectrum to make such comparisons.

If such were the case then we can use the following as an example. Remember that such small shifts as even 2 nm are very unlikely in a variable star. However, perhaps it is instructive for us to understand how we measure such velocities and movements.

Any gaseous movement toward the observer will be shifted to the blue end of the spectrum, as the movement compresses the wavelengths of light, and their frequency shifts to the blue end of the spectrum. Conversely, when gas is moving away from the observer the wavelengths are stretched slightly, letting the frequency decrease so that we have a shift to the red end of the spectrum. These shifts are known as Doppler shifts.

What if we have a star that displays a shift in its wavelength toward the red of 652 nm in comparison to its laboratory wavelength of 650 nm?

$$652 \text{ nm} - 650 \text{ nm} = 2.$$
$$2 \div 652 \text{ nm} = 0.00306 \text{ nm}$$
$$0.00306 \text{ nm} \times 3 \times 10^8 \text{ m} = 918,000 \text{ m/s}$$

To give V_r in km $(918,000 \div 1,000) = 918 \text{ km/s}$

The entire equation also holds for all radial velocities in space as a Doppler or redshift function. If you can find the wavelength of the laboratory spectrum of an element and identify it in the spectra of any astronomical object, one can find its radial velocity in recession or approach.

If the measured wavelength has a greater numerical value than the standard, the object is receding from us (red-shifted). If the measured wavelength is smaller than the standard, the object is moving toward us (blue-shifted). In practice it may be very difficult if not impossible to perform this exercise, as the wavelength shifts of most variable stars is extremely small, and your instrumental resolution may not be up to performing at that scale. Nevertheless, it is instructive to discover the tools that professionals use to make these distance and movement judgments.

Spectroscopy with a DSLR camera

It is possible to perform low-resolution spectroscopy with a DSLR camera, too, though obviously, one will not have the resolution or fine detail that a telescope and CCD camera will provide. However, if you have a DSLR camera fitted with either a standard lens or a fixed focus lens, then you can perform spectroscopy in a limited fashion.

To perform this function, the observer will require a Star Analyzer grating that will fit on the 58 mm thread of the Camera (or equivalent depending on your camera. Canon have a standard kit lens with a 58 mm thread for filters). The Star Analyzer can be modified to fit over a lens cap if it is too small or one can get a step down ring.

Some observers use nothing more than a tripod to gain their image, smearing the star's light into a long trail with the spectrum clearly visible, others use a tracking mount to get the same result. The most important thing is setting the grating so that the image is visible on the camera's LCD screen. This can be done simply by imaging a street light and noting the orientation.

From this point on, it is best to use the system to look at the spectra of 1st magnitude stars and adjust your exposure times so that you gain the maximum image. One can download the image for greater clarity onto a laptop and open the image in software such as Photoshop or RSpec, which is available from https://www.rspec-astro.com. The screen with its resultant spectrum looks like Fig. A.4.

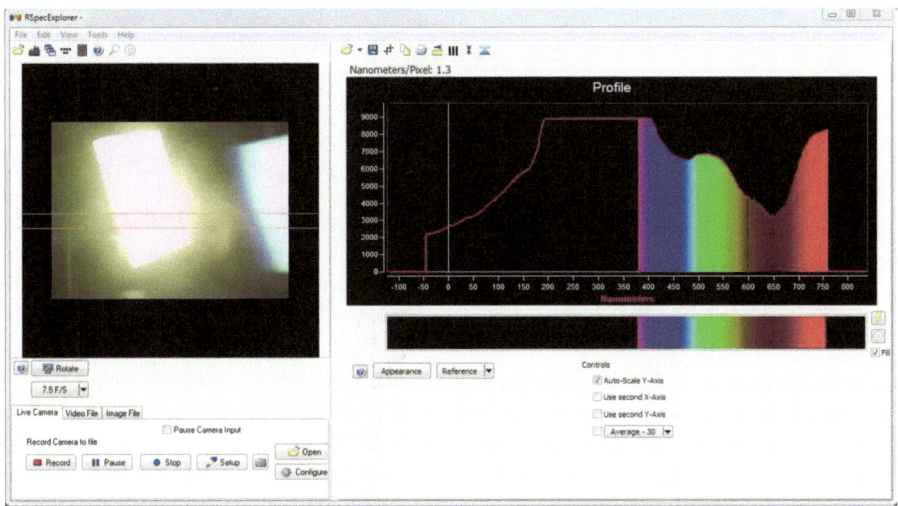

Fig. A.4 RSpec image screen (Image from https://www.rspec-astro.com.)

Once you have your image in this software, orient the spectra so that it is horizontal and then calibrate the spectra in either angstroms or nanometers. There are plenty of "how to" instructions and videos available for tutorials at the above Rspec website.

One can see that spectroscopy is within reach of most experienced amateurs; it all depends on your available instrumentation and how interested you are in performing spectroscopy of variable stars. It is hoped that the foregoing will encourage a few observers to look into this field in far more detail than is covered here, so that an additional, important adjunct to variable star observing can be done and the valuable data collected can be made available to other observers via the AAVSO or the BAA.

Glossary
of Astronomy
Terms

Absolute magnitude   The apparent magnitude that a star would possess it if were placed at a distance of 10 parsecs from Earth. In this way, absolute magnitude provides a direct comparison of the brightness of stars. The apparent magnitude of a star is based upon its luminosity and its distance. If all stars were placed at the same distance then their apparent magnitudes would only be dependent on their luminosities. Thus, absolute magnitudes are true indicators of the amount of light a star emits.

Absorption   Absorption is a property of atomic elements. Atoms absorb a photon of light of a particular wavelength, resulting in the electron(s) within the atom either jumping to a higher orbit in the atom (excitation) or leaving the atom altogether, a process known as ionization. This leads to the development of a dark line in the spectrum of a star or other body at the specific energy or wavelength of the absorbed photon.

Accretion   An accumulation of dust and gas into larger bodies such as stars, planets and moons, or as discs around existing bodies.

Accretion Disc   A disc of material formed around a star such as a white dwarf, drawn from a close binary partner and forming a ring around the equator of the dwarf.

Albedo   A measure of the reflectivity of an object, expressed as the ratio of the amount of light reflected by an object to that of the amount of light incident upon it. A value of 1 represents a perfectly reflecting (white) surface, while a value of zero represents a perfectly absorbing (black) surface. Some typical albedos are: Earth −0.39; the Moon −0.07; Venus −0.59.

Amplitude   The total variation of a light curve from a variable star between maximum and minimum light.

© Springer Nature Switzerland AG 2018
M. Griffiths, *Observer's Guide to Variable Stars*, The Patrick Moore
Practical Astronomy Series, https://doi.org/10.1007/978-3-030-00904-5

Aperture photometry   A system of acquiring light from stars using a CCD camera in order to gain the stars' magnitudes and a host of other information. Usually done in conjunction with software.

Aphelion   The point in an orbit around the Sun at which an object is at its greatest distance from the Sun (opposite of perihelion).

Apoapsis   The point in an orbit when a planet is farthest from any body other than the Sun or Earth.

Apogee   Similar to aphelion. The point in an orbit when a body orbiting Earth (e. g., the Moon or an artificial satellite) is farthest from Earth (opposite of perigee).

Arc minute   A measure of angular separation – one 60th of a degree or one 60th of an hour of right ascension.

Arcsecond   Another measure of angular separation – one 60th of an arc minute. (1/3,600 of a degree).

Ascending node   The point in the orbit of an object when it crosses the ecliptic (or celestial equator) while moving south to north.

Asteroid   (Also "planetoid") These are rocky bodies, the vast majority of which orbit the Sun between Mars and Jupiter. It is thought that there must be around 100,000 in all. The largest asteroid is Ceres, which has a diameter of about 1,000 km. The smallest detected asteroids have diameters of several hundred feet. Together with comets and meteoroids, asteroids make up the minor bodies of the Solar System. They are considered to be the leftover planetesimals from the formation of our Solar System. The gravitational pull of Jupiter is thought to have stopped the members of the Asteroid Belt from forming a planet.

Astronomical unit. (AU)   This is the mean distance from Earth to the Sun, i.e., 149,597,870 km.

Aurora   A glow in Earth's ionosphere caused by the interaction between Earth's magnetic field and charged particles from the Sun (The Solar Wind). It gives rise to the northern lights, or Aurora Borealis, in the northern hemisphere, and the Aurora Australis in the southern hemisphere.

Barycenter   The center of gravity between two massive objects such as stars. They orbit this point in space rather than each other.

Binary star   A system of two stars orbiting around a common center of mass due to their mutual gravity. Binary stars are twins in the sense that they formed together out of the same interstellar cloud.

Broadband filter   A filter that is generally used to reduce light pollution as it transmits the wavelengths of light for Hα, OIII and Hβ but stops the transmission of light wavelengths inimical to sodium and mercury vapor streetlights.

Caldwell catalogue   A catalogue of 110 objects constructed by the British amateur astronomer Patrick Moore and based on the famous Messier catalogue by the 18th century French observer Charles Messier. The Caldwell catalogue is named after Patrick Moore's surname – which was the hyphenated Caldwell-Moore. It contains objects from the NGC and IC catalogues and covers both southern and northern celestial hemispheres.

Cataclysmic variable star A star that suddenly brightens by many magnitudes before fading back into obscurity. Usually dwarf novae type stars or flare stars

CCD camera A camera that uses a charge-coupled device or computer chip sensor to gather light instead of photographic film. It can be used with various colored filters for deep sky photography or for photometry of variable stars.

Celestial equator The projection of Earth's equator upon the celestial sphere.

Celestial sphere The projection of space onto the night sky, an imaginary hollow sphere of infinite radius surrounding Earth but centered on the observer. First postulated by Ptolemy, it is the basis of sky charts and the celestial coordinate system. The coordinate system most commonly used is right ascension and declination. The sphere itself is split up into arbitrary areas known as constellations.

Celestial poles The projection of Earth's poles onto the celestial sphere.

Cepheid variable Cepheid variable stars are intrinsic variables that pulsate in a predictable way. In addition, a Cepheid star's period (how often it pulsates) is directly related to its luminosity or brightness. Cepheid variables are extremely luminous, and very distant ones can be observed and measured. Delta Cephei in the constellation of Cepheus is the most famous of these stars.

Chandrasekhar limit The limit of mass for a white dwarf star, 1.4 solar masses. If this mass is exceeded, the white dwarf will become a type Ia supernova.

Chromosphere The layer between the photosphere and the corona in the atmosphere of the Sun, or any other star, mainly composed of excited hydrogen atoms. In a Hα telescope the chromosphere appears to have a myriad of bright points across the solar disc, a phenomenon known as the chromospheric network. Hα, OIII and Hβ

Cluster Variable A Cepheid variable star with a short period of a day or less. Generally found in globular clusters.

Coma (1) The dust and gas surrounding the nucleus of a comet. (2) A defect in an optical system that gives rise to a blurred, pear-shaped, comet-like image.

Comet An icy object in independent orbit around the Sun, smaller than a planet and usually presenting a highly elliptical orbit extending out to beyond Jupiter.

Comparative photometry The practice of using comparison stars to determine the brightness of a variable star in the same field of view.

Comparison stars These are stars in the same field as a known variable with known magnitudes with which comparisons of the variable star light output and magnitude can be determined.

Conjunction When two bodies appear to be close together in the sky, i.e., they have the same right ascension. Mercury and Venus are said to be at superior conjunction when they are behind the Sun, and at inferior conjunction when they are in front of it. The outer planets are simply said to be in conjunction when they pass behind the Sun.

Constellation An arbitrary grouping of stars that form a pattern. The sky is divided into 88 constellations today. These vary in size and shape from Hydra, the sea monster, which is the largest at 1,303 square degrees, to Crux, the cross, which is the smallest at 68 square degrees.

Corona The outer layer, and hottest part, of the Sun's atmosphere.

Coronagraph A special telescope that blocks light from the Sun's disc, thus creating an artificial eclipse, in order to study its atmosphere.

Cosmic ray An extremely fast, energetic and relativistic (high-speed) charged particle.

Cosmos The universe. The word is derived from the Greek, meaning "everything."

Crater A depression in the lunar or planetary surface caused by an impact from a large meteor or asteroid. Generally circular in appearance and occasionally marked with a central peak and collapsed walls.

Culmination An object is said to culminate when it reaches its highest point in the sky. For northern observers, this occurs when the object is due south. For southern observers when it is due north.

Declination A system for measuring the altitude of a celestial object, expressed as degrees north, or south, of the celestial equator. Angles are positive if a point is north of the celestial equator, and negative if south. It is used in conjunction with right ascension, to locate celestial objects.

Descending node The point in the orbit of an object when it crosses the ecliptic while traveling north to south.

Differential photometry A method of comparing stars and their proper magnitudes by means of analyzing the starlight.

Direct motion (prograde motion) (1) Rotation or orbital motion in an anticlockwise direction when viewed from the north pole of the Sun (i.e., in the same sense as Earth), the opposite of retrograde. (2) The east-west motion of the planets, relative to the background of stars, as seen from Earth.

Digital camera This can be the single lens reflex camera (SLR), which instead of having standard film inside now relies on an imaging chip to capture the scene in the same manner as a video camera or charge coupled device (CCD) camera. It can also refer to any compact digital camera that uses chip technology and to differentiate it from the larger SLR types known as DSLR for shorthand.

DMK camera A camera that uses digital technology to capture image files in the form of a movie that can then be downloaded and stacked in an appropriate software program such as *Registax*. They are used for lunar, solar and planetary imaging.

Dwarf novae Stars that suddenly flare up in brightness for a period less than a day, often for just a few minutes or hours. They are binary stars systems where material is accumulating on a white dwarf star before it undergoes thermonuclear explosion.

Dwarf star A star that lies on the main sequence and is too small to be classified as a giant star or a supergiant star. For example, the Sun is a yellow dwarf star.

Eccentricity The eccentricity of an ellipse (orbit) is the ratio of the distance between its foci and the major axis. The greater the eccentricity, the more 'flattened' is the ellipse.

Eclipse A chance alignment between the Sun, or any other celestial object, and two other celestial objects in which one body blocks the light of the Sun, or other body, from the other. In effect, the outer object moves through the shadow of the inner object.

Eclipsing binary   Eclipsing binary stars are just one of several types of variable stars. These stars appear as a single point of light to an observer, but based on its brightness variation and spectroscopic observations, the single point of light is actually two stars in close orbit around one another. The variations in light intensity from eclipsing binary stars are caused by one star passing in front of the other relative to an observer. The star Algol in the constellation of Perseus is a famous eclipsing variable.

Ecliptic   The apparent path the Sun (and, approximately, that of the planets) as seen against the stars. Since the plane of Earth's equator is inclined at 23.5 degrees to that of its orbit, the ecliptic is inclined to the celestial equator by the same angle. The ecliptic intersects the celestial equator at the two equinoxes.

Ellerman bombs   Microflares in the solar chromosphere associated with magnetic field reconnections, where two opposing streams of ionized material collide with a brief flare of light and energy. A small solar flare.

Elliptical variable stars   Elliptical or ellipsoidal variables are a class of variable star. They are close binary systems whose components are ellipsoidal. They are not eclipsing, but fluctuations in apparent magnitude occur due to changes in the amount of light-emitting area visible to the observer. Typical brightness fluctuations do not exceed 0.1 magnitudes.

Elongation   The angular distance between the Sun and any other Solar System body, usually Earth, expressed in degrees. The term greatest elongation is applied to the inner planets, Mercury and Venus. It is the maximum elongation from the Sun. At greatest elongation, the planet will appear 50% phase.

Emerging flux region   Areas on the Sun where a magnetic dipole, or flux tube, is surfacing on the disc and can produce a bipolar sunspot group.

Ephemeral regions   Limited energy magnetic dipoles with lifetimes of about a day that contain no sunspots. Ephemeral regions can develop anywhere on the Sun, but are more common at mid and lower solar latitudes.

Equatorial mount   A telescope mount designed so that the two axes that support it are aligned, one to the polar axis and the other to Earth's equator. Once an object is centered in the telescope's field of view, only the polar axis need be adjusted to keep the object in view. If the polar axis is driven at sidereal rate, it will counteract the rotation of Earth, keeping the object (except the Moon) stationary in the field of view.

Equinox   This is the time when the Sun crosses the celestial equator. There are two equinoxes: vernal (spring), around March 21, and autumnal (autumn), around September 23. On these dates, day and night are equal. Actual dates and times vary due to Earth's precession.

Eruptive Star   A star in which there is an eruption from the surface, which makes the star vary in brightness.

Faculae   Unusually bright spots, or patches, on the Sun's surface. They precede the appearance of sunspots and can remain for some months afterwards.

Fibrils   Fine structure in sunspot areas associated with spicules and solar activity in the chromosphere.

Field of view   This is the size of the night sky visible through the eyepiece of a telescope or binocular and is measured in degrees, arcminutes or arcseconds.

Filament   A strand of (relatively) cool gas suspended over the Sun (or star) by magnetic fields, which appears dark against the disc of the Sun. A filament on the limb of the Sun seen in emission against the dark sky is called a solar prominence.

Galaxy   A vast star system containing thousands of billions of stars, dust and gas, held together by gravity. Galaxies are the basic building blocks of the universe. There are three main classes, elliptical, spiral and barred, named for their appearance.

Galilean moons   Jupiter's four largest moons: Io, Europa, Ganymede and Callisto. First discovered by Galileo.

Geosynchronous orbit   Sometimes known as a geostationary orbit, in which a satellite's orbital velocity is matched to the rotational velocity of the planet, and as such, a geostationary satellite would appear to be stationary relative to Earth.

Globular cluster   A spherical cluster of older stars, often found around galaxies and containing hundreds of thousands or even millions of stars.

Granulation   The mottled, orange-peel appearance of the Sun's surface, caused by convection within the Sun.

Gun Griz photometric system   A photometric calibration system for professional use that is referenced with known stars of particular spectral character and brightness.

Heliocentric   Sun-centered system of cosmology.

Hump   Fading variable stars will occasionally brighten for a short period of time, giving the associated light curves a characteristic "hump." There are also "super-humps," which last longer than the average period of variability.

Hypersensitize   The process of treating a photographic film with a hydrogen- or nitrogen-forming gas to render the emulsion more sensitive to light and to reduce reciprocity failure with long exposures.

Inclination   (1) The angle between the orbital plane of the orbit of a planet and the ecliptic. (2) The angle between the orbital plane of a satellite and the equatorial plane of the body it orbits.

Inferior conjunction   When Mercury, or Venus, is directly between the Sun and Earth.

Inferior planets   These are the planets Mercury and Venus. They are called inferior planets because their orbits lie between that of Earth and the Sun.

Interstellar medium (ISM)   The material that fill the voids in space between the stars. The ISM is mostly hot vaporous gas such as coronal gas and stellar winds but does contain hydrogen and helium left over from the Big Bang in addition to elements seeded into the ISM by the death of stars in planetary nebulae or supernovae.

Intrinsic variable star   Intrinsic variability is due to physical changes such as eruptions or pulsations in the star itself.

Ionization   The loss of one or more electrons from an atom. In an anion, an atom will gain an extra electron. Ionization plays a role in the Eddington valve or kappa mechanism in Cepheid variable stars.

Julian Day Number   The integer assigned to a whole solar day in the Julian day count starting from noon Universal Time, with Julian day number 0 assigned to the day starting at noon on Monday, January 1, 4713 BC.

Lagrange point   These are points in the orbits of binary systems where the gravitational attraction between each body is canceled out. If a star fills its Roche lobe, materials will tip onto the binary companion through the inner Lagrange point, or L1.

Light curve   Plotting a graph of the difference in magnitude against time of any variable star gives us a typical light curve in which a sinusoidal line may be seen on the graph, detailing the rise and fall in brightness of the star.

Light year   The distance traveled by light in one year, equal to 9.4607^{12} km.

Limb   The outer edge of the disc of a celestial body.

Long-period variable star   The descriptive term long-period variable star refers to various groups of cool luminous pulsating variable stars. It is frequently abbreviated to LPV. The *General Catalogue of Variable Stars* does not define a long-period variable star type, although it does describe Mira variables as long-period variables. The term was first used in the 19th century, before more precise classifications of variable stars, to refer to a group that were known to vary on timescales typically hundreds of days. By the middle of the 20th century, long-period variables were known to be cool giant stars.

Luminosity   Absolute brightness. The total energy radiated into space, per second, by a celestial object such as a star.

Luminence layer   The image taken by a CCD camera through a hydrogen alpha, SII or CaII filter, which is then added to a BVR image to gain maximum input from the astrophysical image.

Lunation   The period between successive new moons.

Magnetic variable stars   These stars have very intense magnetic fields and can be dwarf stars such as highly magnetized white dwarfs or are main sequence stars with high magnetic flux, which cause flares or disruptions to the stellar surface and thus alter the light output.

Magnetosphere   The region of space where a planet's magnetic field dominates that of the solar wind.

Magnitude   The degree of brightness of a celestial body designated on a numerical scale, on which the brightest star has magnitude -1.4 and the faintest star visible to the unaided eye has magnitude 6. A decrease of one unit represents an increase in apparent brightness by a factor of 2.512. Apparent magnitude of a star is the brightness as we see it from Earth, while absolute magnitude is a measure of its intrinsic luminosity. Lower numbers represent brighter objects.

Mare   Areas on the lunar surface that were once thought to be seas of water. (Hence *mare,* Latin for "sea"). Any open surface on a planet that is a lava plain.

Meteor   Also known as a "shooting star" or "falling star," a meteor is a bright streak of light in the sky caused by a meteorite as it burns up in Earth's atmosphere.

Meteorite   A rock of extraterrestrial origin found on Earth

Minor planets   Another term for asteroids.

Mira type   A long-period red giant star with a large variance in magnitude.

Moon   A naturally occurring satellite, or relatively large body, orbiting a planet.

Mylar filter   A solar filter that allows less than 1% transmission of light through a metalized filter to enable safe solar viewing in white light.

Nebula   A term used to describe celestial objects that have a fuzzy, or nebulous, appearance (from the Latin for cloud.), but now used to describe clouds of gas or dust that have condensed out of the interstellar medium (ISM).

Nebula filter   Generally a wide band-pass filter or light pollution filter that allows the passage of Hα, OIII and Hβ wavelengths through to a camera, optical system or CCD camera.

Neutron star   The remnant of a supernovae that has left a core mass of less than 3.5 solar masses behind. Neutron stars are extremely dense and often rotate rapidly, throwing off electrons into their intense magnetic fields and generating radio waves to become pulsars.

Nova   An existing star that suddenly increases its brightness by more than 10 magnitudes and then slowly fades. Novae are generally associated with binary stars in which one of the stars is a white dwarf in close proximity to the primary star. The primary star sheds gas to the white dwarf that allows build up on the surface until pressure and temperature ensure a huge thermonuclear detonation.

Occultation   This is when one celestial body passes in front of, and obscures, another.

Open cluster   A group of young stars, possibly bound together by gravity, that formed together.

Opposition   A planet is said to be "in opposition" when it appears opposite the Sun in the sky. For the outer planets, this is generally the closest they come to Earth, hence when they are most easily visible.

Optical binary   A pair of stars that happen to lie close to one another on the celestial sphere because of a chance alignment. They are not physically associated with one another and lie at vastly different distances. Optical binaries are also known as visual binaries.

Orbit   The path of one body around another due to the influence of gravity.

Orion-type variable   A young variable star with irregular fluctuations in brightness. FUORS are typical.

Parallax   The angular difference in apparent direction of an object seen from two different viewpoints.

Parsec   A unit for expressing large distances. It is the distance at which a star would have a parallax of one arcsecond, equal to 3.2616 light years or 206,265 astronomical units (AU).

Penumbra   Means, literally, "dim light." It most often refers to the outer shadow cast during eclipses, and defines the region of shadow that gives rise to a partial eclipse. It is also the lighter area surrounding the central region of a sunspot.

Periapsis   The point in an orbit closest to a body other than the Sun or Earth.

Perigee   The point in its orbit where the Moon, or planet, is closest to Earth.

Perihelion   The point in its orbit when an object is closest to the Sun.

Perturb   To cause a celestial body to deviate from its predicted orbit, usually under the gravitational influence of another celestial object.

Photometry   The method used to determine the variability of a star over time by using either a photometer or a CCD camera.

Photosphere   The visible surface of the Sun.

Plage   Bright region in the Sun's chromosphere.

Planisphere   An aid to locating stars and constellations in the night sky. It consists of two discs: one with the entire night sky and the other that covers the first, having a window through which a portion of the sky can be seen. The second disc is set according to the date and time.

Precession   Circular motion around the axis of rotation of a body, fixed with respect to the stars. Earth is a giant gyroscope whose axis passes through the North and South Poles, and this axis precesses with a period of 27,700 years.

Prominence   A cloud, or plume, of hot luminous gas in the solar chromosphere. It appears bright when seen against the cool blackness of space. When prominences are in silhouette against the disc they are known as filaments. They are mainly composed of hydrogen, helium and calcium.

Quadrature   When a superior planet – Jupiter, Saturn, etc. – is at right angles to the Sun, as seen from Earth.

Quasars   Compact, extragalactic objects at extreme distances, which are highly luminous. They are thought to be active galactic nuclei. The name is an acronym for quasi-stellar radio source. A quasar is very similar to a QSO (quasi-stellar object) but emits radio waves also.

Radiant   The part of the sky from which a particular meteor stream appears to come. Meteor showers are usually named after the constellation in which the radiant originates.

Range   The amount of change in the brightness of a star—the variation of magnitude from a variable star.

R Coronae Borealis star   Stars at the end of their lives, which suddenly dim by several magnitudes before slowly regaining their original brightness.

Recurrent nova   A star that has undergone more than one outburst, resulting in a large increase in brightness. A cataclysmic variable star.

Red giant   A spectral-type K or M star nearing the end of its life and having a low surface temperature and a large diameter, e.g., Betelgeuse in Orion.

Red shift   The lengthening of the wavelength of electromagnetic radiation caused by relative motion between source and observer. Spectral lines are red-shifted from distant galaxies, indicating that the galaxies are moving away from us due to the expansion of the universe.

Resolution   The amount of small detail visible in an image (usually telescopic), low resolution shows only large features; high resolution shows many small details.

Retrograde   Rotation of a planet, or orbit, opposite to that normally seen.

Right ascension (RA)   The angular distance, measured eastwards, from the vernal equinox. It is one of the ordinates used to reference objects on the celestial sphere and is the equivalent to a longitude reference on Earth. There are 24 hours of right ascension within 360 degrees, so one hour is equivalent to 15 degrees.

Together with declination, it represents the most commonly used coordinate system in modern astronomy.

Semi-major axis   The semi-major axis of an ellipse (e.g., a planetary orbit) is 1/2 the length of the major axis, which is a segment of a line passing through the foci of the ellipse, with end points on the ellipse itself. The semi-major axis of a planetary orbit is also the average distance from the planet to its primary.

Semi-regular variable   A star that varies in brightness over an undetermined time but may show some regularity in its light curve so that it can be classed as almost or semi-regular.

Sidereal time   Star time; the hour angle of the vernal equinox. Time measured with respect to the fixed stars rather than the Sun.

Sidereal month   The 27.32166-day period of the Moon's orbit.

Solar continuum filter   A green light filter transmitting light wavelengths centered at 510 nm, rendering a visible green image of the Sun. Such filters are used in conjunction with either a Herschel wedge or Baader astro filters.

Solar cycle   The 11-year variation in sunspot activity.

Solar flare   A sudden, short-lived burst of energy on the Sun's surface, lasting from minutes to hours.

Solar wind   A stream of charged particles emitted from the Sun that travels into space along lines of magnetic flux.

Solstice   This is the time when the Sun reaches its most northerly or southerly point (around June 21 and December 22, respectively). It marks the beginning of summer and winter in the northern hemisphere, and the opposite in the southern hemisphere.

Spectral classification   A method of classifying stars based upon the appearance of hydrogen absorption lines in their spectra. The spectral sequence OBAFGKM was determined by Williamina Fleming and Annie Jump Canon in the early 20th century.

Star cluster   A loose association of stars within the Milky Way. Examples are the Pleiades (Seven Sisters) and Hyades clusters.

Sunspot   A cooler region of the Sun's photosphere (which, thus, appears dark) seen as a spot on the Sun's disc. Sunspots are caused by concentrations of magnetic flux, typically occurring in groups or clusters. The number of sunspots varies according to the Sun's 11-year cycle. More sunspots are seen at the Maxima of solar cycles, with few being observed during the minima between.

Superior conjunction   This is when Mercury, or Venus, is behind the Sun.

Superior planets   Also known as the outer planets. These are the planets beyond the Earth's orbit. They are, in order: Mars, Jupiter, Saturn, Uranus, and Neptune.

Supernova   An exploding star, usually quite massive in comparison to the Sun. There are two main types of supernova: Type Ia are white dwarf stars that exceed the Chandrasekhar mass of 1.4 times that of the Sun, while Type 1B Ic and Type II are massive stars that explode once the iron fusion stage is reached.

Symbiotic star   A star that shares material with a binary star companion in a close orbital system.

T association A collection of young stars that have recently formed from a cloud of gas and dust or are still entwined in their nebula.

Terminator The boundary between day and night regions of the moon's, or a planet's, disc.

ToUcam A small webcam that fits in the eyepiece holder of a telescope to gain a direct video image of an astronomical object. Manufactured by Phillips.

Transit The apparent journey of Mercury or Venus across the Sun's disc, or of a planet's moon across the disc of its parent.

UBVRI The colored filter photometric system generally employed by amateur astronomers and systematized by Michael Bessell in the 1990s, taken from original work by Johnson and Cousins in the 1950s and 1960s.

Umbra From the Latin for shade, it is the shadow area defining a total eclipse. Also the dark central region of a sunspot.

Unsharp masking A photographic and image reduction technique that allows the stacking of many images to gain increased detail and resolution in an astronomical object.

Variable star Any star whose brightness or magnitude varies with time. The variations can be intrinsic because of internal processes, or extrinsic, due to eclipses, dust and other phenomena. Variations can also be irregular or periodic.

White dwarf A whitish star, of up to 1.4 solar masses, and about the size of Earth with consequential very high density, characterized by a high surface temperature and low brightness.

Wratten filters Colored glass filters with a range across the visible spectrum from red to blue that enable the blocking of particular long-pass wavelengths of light in order to see more detail on planetary and lunar surfaces. The filters are indicated by particular numbers that are standardized across a range of colors.

Zenith The point on the celestial sphere directly above an observer, or the highest point in the sky reached by a celestial body.

Zenithal hourly rate (ZHR) The number of meteors per hour, for a particular meteor stream, that is estimated will be seen under favorable seeing conditions if the radiant were directly over the observer. Usually the actual figure is less than this.

Zirin class The different active or quiescent features of prominences in the solar chromosphere, developed by Harold Zirin.

Zodiac The apparent path, in the sky, followed by the Sun, Moon and most planets, lying within 10 degrees of the celestial equator. Ancient astrologers (nothing to do with modern astronomy!) divided it into 12 groups, the signs of the zodiac, though there are actually 13 astronomical constellations that lie on the zodiac, since the Sun passes through Ophiuchus each December. Ophiuchus is not recognized by astrologers.

Zodiacal light A faint glow from light scattered off interplanetary dust in the plane of the ecliptic.

References
and Further
Reading

AAVSO *The AAVSO Manual for Visual Observing of Variable Stars* AAVSO, Cambridge Mass. 2013

AAVSO *Variable Star Type Designations in VSX* AAVSO Cambridge Mass. 2014

BAA *Observing Guide to Variable Stars* British Astronomical Association, London 2015

Burnham R. *Burnham's Celestial Handbook* (3 Vol) Dover Publishing New York 1978

Durkin M. *DSLR Photometry*. BAA guide https://www.britastro.org/vss/DSLR_PHOTOMETRY.pdf

Good J. *Variable Stars*. Springer Press 2009

Glasby John. *Variable Stars* Constable Books 1989

Holland K. *Why Observe Variable Stars* BAA Workshop, Cambridge Institute of Astronomy 2015. http://www.britastro.org/vss/holland.pdf

Hoot J. *Photometry with DSLR Cameras*. The Society for Astronomical Sciences 26th Annual Symposium on Telescope Science. Journal of the Society for Astronomical Sciences.

Jones K G (ed) *Webb Society Deep Sky Observers Handbook Volume 8 – Variable Stars* Webb Society 1990

Kallrath J. *Eclipsing Binary Stars. Modelling and Analysis* Springer-Verlag 1999

Levy David. *Guide to Variable Stars* CUP 2005

Littlefield C. et al AAVSO DSLR Observing Manual: https://www.aavso.org/dslr-observing-manual

Loughney D. *Variable Star Photometry with a DSLR* Camera BAA Journal 2010. http://www.britastro.org/vss/JBAA%20120-3%20%20Loughney.pdf

Morris S.L. *The Ellipsoidal Variable Stars* Astrophysical Journal. August 1985

Nicholson Martin. *Discover Your Own Variable Star* CUP 2012

North G. *Observing Variable Stars, Novae and Supernovae* CUP 2014

Percy J. *Understanding Variable Stars* CUP 2011

© Springer Nature Switzerland AG 2018

M. Griffiths, *Observer's Guide to Variable Stars*, The Patrick Moore
Practical Astronomy Series, https://doi.org/10.1007/978-3-030-00904-5

Petit Michael. *Variable Stars* Blackwell 1987

Sherwood V. *Variable Stars and Stellar Evolution* IAU Press 2007

Sterken C. & Jaschek C. *Light Curves of Variable Stars* CUP 2008

Upton Winslow. *Star Atlas Containing Stars Visible to the Naked Eye and Clusters, Nebulae and Double Stars Visible in Small Telescopes, Together with Variable Stars.* Forgotten Books 2018

Vermeriss W. *Cataclysmic Variable Stars, How and Why They Vary* Springer 2001

Warner B. *Cataclysmic Variable Stars* CUP 2008

Williams T & Saladyga M. *Advancing Variable Star Astronomy: The Centennial History of the American Association of Variable Star Observers* CUP 2011

Index

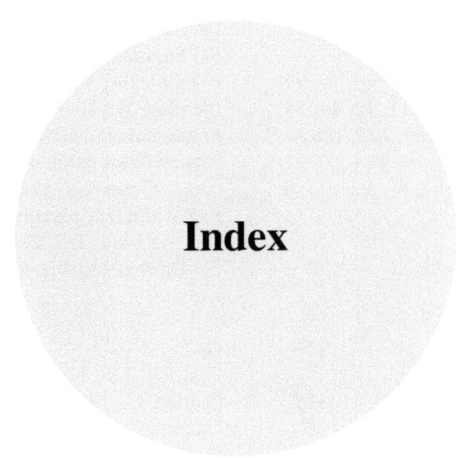

© Springer Nature Switzerland AG 2018   313
M. Griffiths, *Observer's Guide to Variable Stars*, The Patrick Moore
Practical Astronomy Series, https://doi.org/10.1007/978-3-030-00904-5